Risk

The Decision Matrix

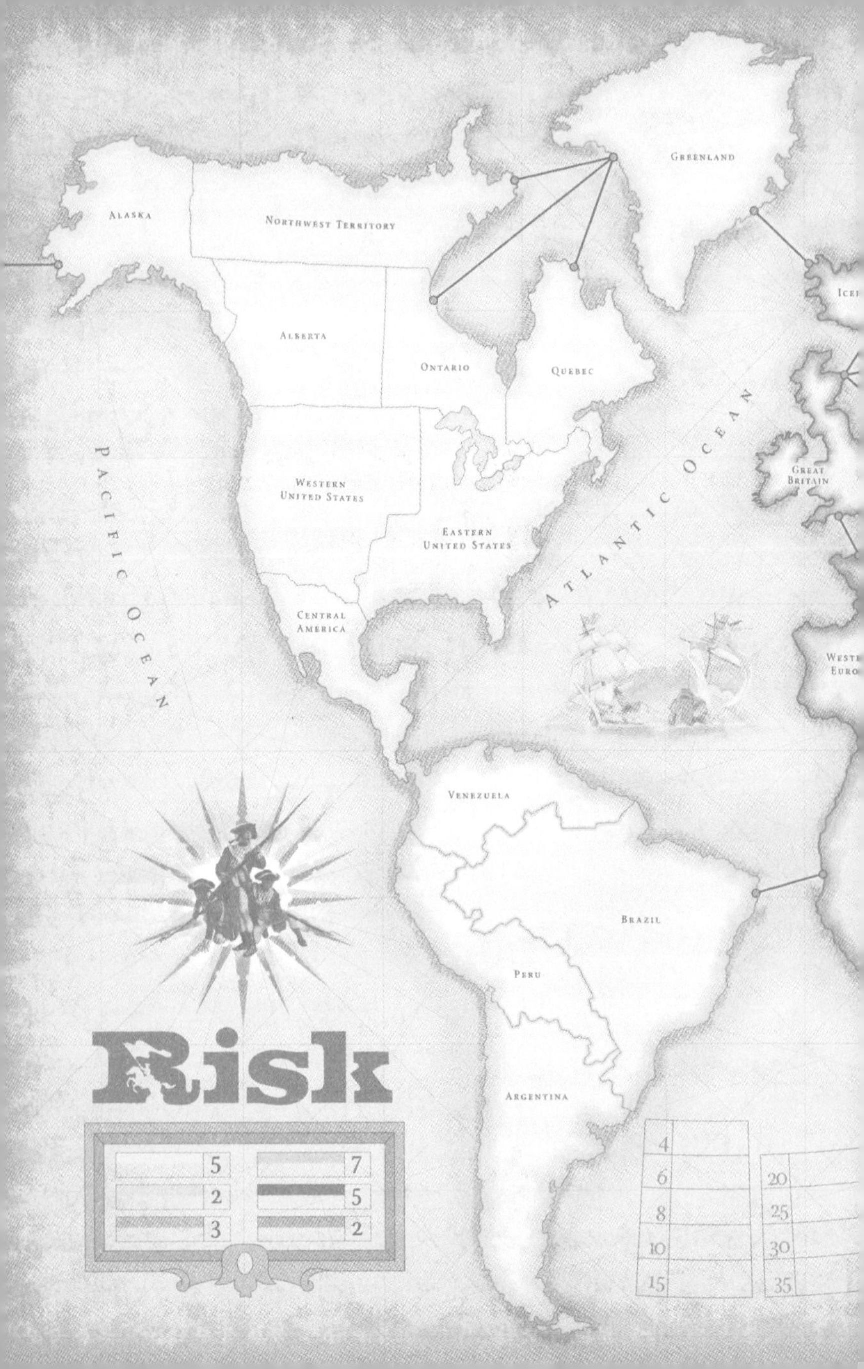

Alaska
Northwest Territory
Greenland
Alberta
Ontario
Quebec
Western United States
Eastern United States
Central America
Pacific Ocean
Atlantic Ocean
Great Britain
Venezuela
Brazil
Peru
Argentina
Risk
5
2
3
7
5
2
4
6
8
10
15
20
25
30
35

Risk

The Decision Matrix

STRATEGIES THAT WIN

ALAN AXELROD

STERLING
New York / London
www.sterlingpublishing.com

Library of Congress Cataloging-in-Publication Data

Axelrod, Alan, 1952-
RISK, the decision matrix : strategies that win / Alan Axelrod.
p. cm.
Includes bibliographical references and index.
ISBN 978-1-4027-5410-4
1. Risk–Sociological aspects. 2. Strategy. 3. Game theory. 4. United States–History–20th century. I. Title.
HM1101.A94 2008
302'.12–dc22

2008032431

10 9 8 7 6 5 4 3 2 1

Published by Sterling Publishing Co., Inc.
387 Park Avenue South, New York, NY 10016

Distributed in Canada by Sterling Publishing
c/o Canadian Manda Group, 165 Dufferin Street
Toronto, Ontario, Canada M6K 3H6
Distributed in the United Kingdom by GMC Distribution Services
Castle Place, 166 High Street, Lewes, East Sussex, England BN7 1XU
Distributed in Australia by Capricorn Link (Australia) Pty. Ltd.
P.O. Box 704, Windsor, NSW 2756, Australia

Book design and layout: Sherry Williams/Oxygen Design
Please see picture credits page for image credits.

Manufactured in the United States of America

Sterling ISBN 978-1-4027-5410-4

For information about custom editions, special sales, premium and corporate purchases, please contact Sterling Special Sales Department at 800-805-5489 or specialsales@sterlingpublishing.com.

CONTENTS

INTRODUCTION

RISK as a Model for Modern Decision Makers

Arthur Conolly was a man of many guises. An explorer and writer, he was also a captain of the Anglo-Indian 6th Bengal Light Cavalry, but functioned more immediately in the employ of the British East India Company as a spy. Intrepid and bold, he took espionage seriously, yet also with a sufficient sense of humor to travel incognito throughout Central Asia under the name Khan Ali, a Middle Eastern play on his distinctly Western surname. Writing in 1829 to Sir Henry Rawlinson, another man of many works–soldier, diplomat, and gentleman-scholar of ancient Persian texts–Conolly coined the term the "Great Game" as a label for the diplomatic contest between the British and Russian empires for political control over Central Asia.

For Conolly, the Great Game ended in June 1842 when Nasrullah Khan, the emir of Bukhara, executed him as a British spy. But his phrase lived on. Historians came to use it to describe the intense period of diplomatic maneuvering between the two empires spanning roughly 1813 to 1907, but it was also applied to global diplomacy more generally thereafter, up to the period of World War II.

Following that world-consuming conflict, some historians recognized the start of the "New Great Game," a label they bestowed on the Cold War jockeying for global dominance—no longer between the British and the Russians, but between the Americans and the Soviets.

If strategy is an overall plan for accomplishing some goal—whether the strategy is produced by and applied to a single person, a business enterprise, or an entire nation—diplomacy is the omnibus term for all the means, moral or immoral, peaceful or warlike, used to engage with other people, enterprises, or nations to get what you, your enterprise, or your nation wants. Whatever the scale involved—individual, corporate, or national—diplomacy is deadly serious, with everything at stake. Yet Conolly and many others after him saw it all as a game.

Cynical? Shocking? Or just a little backwards? We are, after all, accustomed to children imitating "real life" in the games they play. The surprise is that real life—serious and consequential though it assuredly is—can often imitate a game. The perception, or perhaps the realization, that the relationship between games and life is not one way, but interchangeable in direction, is behind the current interest in game theory among politicians, diplomats, military heads, and business leaders.

Once you accept the two-way flow of game into life and life into game, the notion of a "Great Game" no longer seems so shocking, cynical, irreverent, or trivial. Only consider that game theory was born in 1944 when the eminent mathematician John von Neumann teamed up with the economist Oskar Morgenstern to write *Theory of Games and Economic Behavior*, which soon became the basic textbook that the legendary Cold War–era think tank, the RAND Corporation, used to develop the scenarios and strategies that formed the basis of America's nuclear weapons policy. In the Atomic Age, playing at diplomacy became playing at doomsday. And, strange as it sounds, that was a good thing, since it is far better to play at doomsday than actually to implement it.

Whether applied to global thermonuclear war, Cold War diplomacy, current diplomacy, strategies for big businesses and small, or to individual decision making, game theory is a means of modeling and studying strategic interactions between players (or "agents"—the people who make decisions and take actions). It is fundamental to game theory that players choose strategies they believe will maximize their return, always in the context of the strategies chosen by the *other* players of the game. Game theory assumes that, in "social situations" (situations that may be on the scale of just two or three individuals, the entire planet, or somewhere in between), people ("agents," "CEOs," "national leaders," "generals," call them what you will) likewise attempt to embrace strategies of optimization.

In other words, everyone tries to win.

And since everyone in a given social situation is trying to win, the strategies they choose interact, compete, conflict, cooperate, succeed, and fail, always in relation to one another. There are many variables that make the study of comparative strategy in social situations both fascinating and, typically, desperately complex. For example, human variables include rationality, irrationality, genius, wisdom, skill, mediocrity, incompetence, misperception, idiocy, and the like. Natural variables may encompass climate, weather, topography, and population. Add to this a handful of random variables, including luck, chance, destiny, and Divine Providence.

Because the real world presents so many variables, modeling strategy within the far simpler spatial and temporal confines of a game can be a very valuable way of understanding and evaluating different strategies, pitting one against another. Equally important is the fact that it is almost always far more economical to "game out" a personal life choice, a business strategy, or a proposed national policy than it is to jump into, say, a marriage, launch a new product line, or invade a nation—all for real—and simply hope for the best.

Game theorists have used many different games to model the reality of social situations, ranging from the most basic—simple card and dice games—to the most ancient—chess and go, for example—to more modern games. (My own *Everything I Know about Business I Learned from Monopoly* was published in 2002.) But no contemporary game holds greater interest and more potential for game theory—for modeling real-world strategy—than RISK.

The game was originally the brainchild of Albert Lamorisse, best known to film buffs as the writer, director, and producer of the classic 1956 French short film *The Red Balloon,* but immortalized among game enthusiasts as the creator of "La Conquête du Monde"—World Conquest—a board game published in France in 1957 by a company called Miro. That same year, Miro approached the American game-making giant Parker

Brothers, which purchased "La Conquête," made some revisions to the basic rules, and in 1959 brought out "Risk: The Continental Game." Some believe the new name was a stroke of genius. Others say that it represented nothing more than the initials of the four grandchildren of the Parker Brothers executive in charge of marketing the game. In either case, the RISK title was soon tweaked a final time as "RISK: THE GAME OF GLOBAL DOMINATION."

Since its introduction, RISK has appeared in various additional versions. There have been *Castle Risk* (1986), *Risk: Édition Napoléon* (1999), *Risk: Édition Napoléon: Extension Empire Ottoman* (2000), *Risk: 2210 A.D.* (2001), *Risk: The Lord of the Rings* (2002), *Risk: The Lord of the Rings: Gondor & Mordor Expansion Set* (2003), *Risk: The Lord of the Rings: Trilogy Edition* (2003), *Risk: Godstorm* (2004), *Risk: Star Wars: Clone Wars Edition*, *Risk: Star Wars Original Trilogy Edition* (2006), *Risk Junior: Narnia* (2006), and *Risk: The Transformers Edition* (2007), which is based on the *Transformers* blockbuster movie. In addition to the board versions of the game, there are computer-based and online formats. Parker Brothers and Avalon Hill, which publish the various versions of the game, are now divisions of Hasbro.

In its original form and format, RISK is played by two to six players, one of whom survives to the end of the game and is therefore the winner: the player who has conquered the world. There are no ties and no ambiguities. At the end of the game, the sole conqueror remains. Every token on the board is the same color and belongs to the victor.

The RISK game board is a simplified world map divided into 42 territories, and the game comes with six sets of armies, each of a different color, with each set consisting of infantry, cavalry, and artillery pieces. There are also five dice, which are used to determine the outcome of attacks and defenses, and cards, which are collected as a player conquers territories and which are used to secure reinforcements to the player's army. The object of the game is simple: to win by conquering the world. This means acquiring territories and continents by using one's army tokens to occupy unoccupied territories and by successfully invading territories held by others, yet without leaving oneself vulnerable to invasion and conquest in return. In each of the game's many turns—and games often assume marathon proportions—each player must decide whether to attack, where to attack, when to attack, and when to stop attacking.

That is the game in a proverbial nutshell. But because this description leaves unanswered the question of why RISK is so interesting to game theorists who want to model the real world realistically, we need to push the discussion a bit further.

Unlike chess, which models decision making on a highly logical, even quasi-mathematical level, requiring the ideal player to think strictly in terms of perfect moves, RISK provides a multiplicity of winning (and losing!) options with every turn in each game. There is no one right or wrong strategy. There is no one right or wrong move. Strategic and tactical choices alike offer a whole range of benefits and liabilities.

In RISK, spatial strategy is important—placing pieces for effective occupation, conquest, defense, and movement from one territory to another—but so is the ability to envision, formulate, and execute genuine negotiations with other players. As in poker played at the highest of levels, a capacity to "read" others, to see the game from their perspective, to get into their heads, to manipulate them, to cajole them, to persuade them, and even to deceive them, is critically important to victory. Yet also as in poker—and in contrast to chess—chance plays a significant role, just as it does in real life. The winning player cannot rely solely on chance, of course, but must always factor it into strategy and style of play.

This book, like its companions in the RISK series, is not a how-to manual for winning RISK, and you don't even need to play RISK to enjoy and benefit from the book. In fact, we won't even be talking specifically about the game of RISK in the rest of the book. The chapters that follow are devoted instead to some of the most significant—that is, momentous and instructive—strategic decisions in history: the game of "real life." And that game gets you deep into the same issues RISK offers in model form only:

- A competition for limited resources
- The absolute necessity of making decisions
- The necessity—if you hope to win—of making strategic decisions
- The ever-present potential for conflict
- The ever-present potential for the resolution of conflict through war or diplomacy

- The need to define, create, and end relationships, including alliances and enmities
- The need to influence, persuade, dissuade, intimidate, force, and cooperate with others
- The necessity for managing time and timing
- And the urgent requirement that chance be accepted, managed, exploited, coped with, and compensated for

As a rich example of game theory, RISK is neither more nor less than a stunningly useful way to start thinking clearly about strategic decision making in the real world—the game board on which the chapters of this book will be played out. The object? To learn to make the moves that will win the greatest "Great Game" of them all: *your* profession, *your* passion, *your* life.

CHAPTER 1

Visualizing the Decision

The Art of Putting Together the "Big Picture"

Previous page: ***RISK game board.***

Visualizing the Decision

ONE OF THE MOST APPEALING THINGS about the classic board games is, quite simply, *the board.* As with its older cousins, chess, Go, and even checkers, playing RISK requires deploying pieces on a board. The decisions involved in this deployment require tactical, logistical, and strategic thought—that is, decision making on an intimate, move-by-move scale (tactics) as well as the larger scale of the overall "big picture" (strategy), with logistics occupying a realm of decision making somewhere between the micro level of tactics and the macro level of strategy. (In the case of RISK, this mainly involves maintaining the ability to move and maneuver, whether to attack, occupy, or withdraw.) To a degree far greater than in chess, checkers, or Go, RISK adds to tactics, strategy, and logistics another dimension, diplomacy, which is based on negotiation.

Diplomacy, the interpersonal level of RISK, is not itself reflected directly on the board but in the actions carried out on the board, including movement, invasions, attacks, and the like. Thus, even with its additional invisible dimension of diplomacy, RISK, like other classic board games, makes decisions visible on the board and requires visualizing relationships, deploying the game pieces to make winning moves.

In the game that we call the "real world," visualizing the impact of our decisions is also critical to winning because decision making always has an impact on relationships, whether between and among people, companies, or nations. Too often, we think of decision making as something we must do alone, for ourselves. The harder the decision, the higher the stakes, the more self-involved we usually become, perhaps even closeting ourselves behind closed doors so that we can be alone to "think this thing through." There is no denying that concentration is important in making good decisions, but it is a fatal error to equate the solitude that may help you to concentrate with the nature of the decision that must be made. No winning decision is ever made in isolation. All meaningful decisions are about you in relation to others—other people, other things, other factors. They are never about you alone.

Think about it. If what you did never affected anyone but yourself, and if what you could do was never affected by anyone or anything outside of yourself, there would be no need for making decisions at all. Whether you *thought* about doing X or Y or whether you simply drifted *thoughtlessly* into doing X or Y would make no difference, since there would be nothing to limit or prompt your choice. With no one and nothing else impacting or impacted by your decision, you could always change your mind. It would be rather like playing RISK solo. No matter where you moved or how long it took, you would eventually occupy all of the territories.

Once we accept that decision making is always about relationships, the advantages of visualizing those relationships in advance of actually making a decision become quite apparent.

But how do you do it? How do you *see* relationships?

On a game board, it is relatively easy—or at least conceptually straightforward—to visualize relationships and the impact various decisions might have on them. All you need to do is take a snapshot of the present and imagine what the future will look like if you decide A, what it will look like if you decide B, and what it will look like if you decide C. In life beyond the game board, visualizing becomes far more metaphorical than literal. People, money, and things are not, after all, so many RISK tokens to be moved here and there across a flat rectangular space measuring roughly twenty by thirty inches. In life, there are more than the two dimensions of plane geometry—and more, even, than the three of solid geometry. Indeed, the most important dimensions in which the moves of real life take place are not even physical, flat, or solid; instead, they occupy spaces of mind and emotion and time. To visualize decisions in this kind of "space" requires a different kind of mapping, a mental mapping that has sometimes been called the "Five Ps Rule."

It works like this. Each of the following Five Ps is a dimension in which any real-world decision, if acted upon, will have to exist, move, and interrelate with the actions of others.

The Five Ps

1. **Problem.** When you approach a decision, ask—and answer—the question: what problems are you trying to solve? It helps to divide the concerns you identify into short- and long-term problems. Short-term problems include those issues and obstacles that need to be addressed more or less urgently as well as those things you must do, overcome, complete, or achieve in order to move toward solving your long-term problems. This second category of short-term problems (immediate steps toward solving long-term problems) require good tactical decisions, decisions that not only solve each short-term problem but also contribute toward solving your long-term problem: achieving your strategic goal.

2. **Position.** Know who you are and where you are in relation to others. The Delphic Oracle of ancient Greece had only one piece of truly salient wisdom to offer: know thyself. The trick to doing that is to look both within and without. You must define who you are in an inner dimension as well as an outer one. Alone, neither of these dimensions is sufficient to self-understanding. To know who you are, you need to decide who you are in relation to others. Evaluating the dimension of "position" requires an honest self-inventory and self-appraisal, including a frank assessment of your strengths and weaknesses, your triumphs and your mistakes.

3. **Possibilities.** Surveyors, navigators, and astronomers all measure distances using triangulation: applying the known location of two points to determine the unknown location of a third point. In the mental space of real-world decision making, you can do much the same thing. If you have assessed your Problem(s) and your Position, you can go on to the third leg of the triangle: Possibilities—options, choices, available moves. Assuming the decision at issue involves sufficiently high stakes to be worthwhile, more than one possibility will emerge from your triangulation. Evaluate each possibility in terms of its comparative risks versus benefits, its cost versus reward. The principle of scarcity, the central concept of the modern science of economics, looms over any meaningful choice we have to make. It is a fact of life that none of us ever has sufficient resources to do everything. If we did, it would never be necessary to make decisions. The reality is that we have resources sufficient to do a limited number of things, and so we must do a trade-off analysis for each decision that has to be made. If I decide to do A, I cannot do B. Or: If I decide to do B, I must give up C.

4. **Proposal.** After you have triangulated your Possibilities based on your Problem and your Position, you are ready to proceed to the fourth P: a Proposal for action. Although each of the first three Ps required thinking about yourself

in relation to whom you are, what you want, and what your position is in relation to others, formulating the Proposal requires even wider and deeper consideration of others. Having decided on a Possibility, your Proposal must next spell out the means by which you will transform the Possibility into reality. How will you bring this idea into the world? Once you have resolved to create—to bring something into the world—you have made a decision to change the world in some way. Acting on your Proposal will impact others, who will, in turn, impact you. Acting on your Proposal will solve your Problem (perhaps), but it will doubtless create other Problems. It will change your Position in the world for the better (perhaps), but it will result in a new Position, which you will have to accommodate. It will transform a Possibility or a set of Possibilities into a new reality (perhaps), but that new reality will create new Possibilities while also eliminating some others. Building the Proposal requires envisioning as many dimensions as possible of the new reality that will be created when you implement your decision.

5. **Price.** Only after you have crafted a Proposal—which includes the choice of Possibility or Possibilities plus the means of implementation—can you finally predict and evaluate the cost of the decision. Cost has more than one dimension. Cost can be measured in money, material, human resources, time, and labor. Often, cost also includes an emotional or spiritual dimension: anxiety, fear, assault on your self-image (for example, Are you really willing to appear ruthless?), sacrifice of friendship, and so on. The Price, moreover, may be paid in the present, in the future, or both. Business planners, for example, often speak of "opportunity costs." Your decision not to make a huge capital investment in the new machine tools to manufacture Product A saves you $XXX this year, significantly adding to your bottom line, but your decision not to buy the new equipment also means that your competitor will capture the entire market for Product A, an "opportunity cost" to you of $XX each year beginning five years from now.

The Five Ps are an aid to orienting you in the multidimensional space of real life, a "game" beyond the confines of a two-dimensional board. The Five Ps should help you visualize your problem, your position, and your possibilities in relation to other "players"; however, this method of visualization is still deeply rooted in your own particular perspective. The Five Ps are mostly about predicting what decisions are possible and what decisions are best, including how they will impact others and how others will impact them. Imagine the realm of real-world decision making as a deep swimming pool. The Five Ps will get your face into the water, open your eyes, and let you look around. But how do you take the plunge? How do you get into the swim?

The answer is both simple and hard.

After the Five Ps have given you all they can, stop thinking about what *you* want and need, and start thinking about what *everyone else* wants and needs. Start thinking about this full time, all of the time.

Consider the last automobile salesperson who succeeded in selling you a car. How did she do it? Did she explain *her* wants and needs to you?

"Look, you have to buy this car, because I need the commission to pay my mortgage."

Certainly not.

Instead, she addressed *your* wants and needs. Your basic need, of course, was transportation. But everyone knows that nobody buys a car just for transportation. If she was a good salesperson, she asked questions in order to ascertain your wants and needs. You *want* a powerful, sexy machine, but you *need* enough space for a spouse, three kids, and a dog. You *want* a Lamborghini, but you *need* it at the price of a Hyundai. Somewhere between *your* wants and *your* needs, she found a car likely to appeal to *you*. She knew what she wanted and needed, but by focusing instead on what you wanted and needed, she made the sale—and, incidentally, by collecting her commission, was able to satisfy her own wants and needs.

This last point is crucially important. Thinking about others' wants and needs does not mean sacrificing your own. On the contrary, it is the only reliable means of satisfying them. By knowing and addressing what others want and need, you gain a significant measure of influence and control over others. To the degree that you influence and control others, you can create a world that is friendly to the implementation of your decisions.

Think of it this way. Four people, including you, are playing RISK. You make the best decisions; therefore, you win. Or: You manage to manipulate the three other players to make decisions that benefit you; therefore, you win. In both cases, the result is the same—but in the second instance, you have managed to get the others to do all the heavy lifting.

Make Them Make You the Winner

It is always a sound decision-making strategy to influence others to create the conditions conducive to your success. How do you implement such a strategy? Consider using the following tactics:

1. **An appeal to reason.** Use facts and logic to persuade others to take a certain course of action. Demonstrate in factual detail how doing X will satisfy the wants and needs of the other person.

2. **An appeal to imagination.** Paint a vivid picture of the good things that will happen to the other person if he does what you want him to. "Imagine what it would be like for us if we partnered on this deal! The market is young, growing, and ready to be entered. We will have a genuine ground-floor advantage."

3. **Ask questions.** The world's best salespeople punctuate more of their sentences with question marks than with periods or exclamation points. Ask questions:

   Q: "Do you want to make a lot of money?"

   A: "Yes."

   Q: "Do you want to make money with minimal risk?"

   A: "Yes, of course I do."

   Q: "If you had a partner who knew the market, would that help you make money and lower your risk?"

   A: "I suppose it would."

Pitch: "Well, here's the deal . . ."

Asking questions elicits answers, each of which represents the other person's investment in your idea as he becomes increasingly engaged in the reasoning process. Moderately persuasive people succeed in persuading others. Masters of persuasion succeed in getting others to persuade themselves.

4. **Offer a deal.** There are two ways to get what you want from someone: theft or business. The thief takes value. The businessperson exchanges value for value. He makes a deal, offering something in return for something else. This kind of transaction is so basic, so paradigmatic that it is hard to resist. Our civilization is built on it: "If you give me X, I will give you Y."

5. **Wheel in the bandwagon.** Use precedent, a trend, or the viewpoint of others to exert influence on someone. "George, sooner or later *everyone* does business with me. Just ask around!"

6. **Assert policy.** Base your appeal on an assertion of authority—a rule, principle, or policy. For better or worse, big companies do this all of the time. "I can't give you a discount. It's against company policy. My hands are tied." This tactic is somewhat coercive or, at least, inflexible. Joe is reluctant to take the deal you offer. "Joe, I have to tell you that my policy is simple. Once someone refuses to make a deal with me, I don't offer a second time." Although this tactic does not present a logical argument for a given course of action, but simply asserts the necessity of abiding by "policy" or "principle" or "the rules," it can be quite effective, provided that you are willing to give an arm a twist and the other person is likely to surrender or at least compromise.

7. **Make a threat.** Flex your muscle. Explain the consequences of failing to cooperate. Tribal and national leaders have been doing this since time immemorial, issuing threats that range from, "If you don't do X, we will curtail trade with you" to "Do X today or we begin bombing tomorrow." This tactic is significantly more coercive than asserting policy. A threat is a last resort, but that is no reason to reject the prospect of making one. Do understand, however, that to make your threat credible, you either need to control a sufficient degree of force or possess the ability to make a highly believable bluff.

Strategically addressing the needs and wants of others opens up all dimensions of the intellectual and emotional space in which decisions are made, but once you enter that space, you will find that it is highly dynamic. The people you must persuade have needs and wants, to be sure, but they are not likely to be static and unchanging. Ultimately, making and implementing winning decisions requires self-evaluation, including an analysis of all five Ps, then persuading others to create an environment friendly to your decisions, and, finally, actively negotiating with those others. The fact is that, in the real world—as in a sophisticated game such as RISK—the end phase of much decision making is negotiation, the fulfillment of your own needs and wants in exchange for at least apparently fulfilling those of others.

As in any other phase of decision making, acquisition of accurate information is key to a successful negotiation. The person who understands what he himself needs and wants as well as what the other person needs and wants is the more likely to come out ahead as a result of the exchange. Decisions made and negotiations conducted in any degree of ignorance, however, are likely to prove unsatisfactory, if not downright disastrous.

Skilled negotiators find it most useful to systematize their acquisition of information by defining for themselves a "settlement range," the range of all possible negotiation outcomes ("settlements") deemed acceptable. At the lowest end of the settlement range is the "least acceptable settlement," the minimum result you are willing to accept. At the highest end is not a pie-in-the-sky fantasy or a dream-come-true result (*I get everything. You get nothing.*), but what students of negotiation call the "maximum supportable settlement," the most you reasonably believe you can get. This maximum point is where you begin the negotiation, albeit with an unspoken willingness to work downward, if necessary, closer to (but never below) your least acceptable settlement.

In negotiation, getting emotional is usually not productive because your feelings get in the way of the hard information that should go into formulating the low and high points of the settlement range. Consider this simple example: You are selling your high-end mountain bike. Ten years ago, you paid $4,500 for it. You have maintained it impeccably, so its mechanical and cosmetic condition is excellent. You would love to get $3,500 for it, but Internet research suggests a going price range from $1,000 to $3,000. You are not willing to part with the bike for $1,000, but, based on your research, you also decide to abandon your *dream of* $3,500 as unrealistic—or above the maximum *supportable* settlement point. So you set your least acceptable settlement figure at $2,000 and your maximum supportable settlement figure at $3,000, the upper limit of what the bike goes for on the Internet. Now you can walk into the negotiation armed with a good map, and you can make your final decision on the basis of real information.

The secret to establishing useful minimums and maximums is to replace your emotions with as much hard information as you can get. Once this is done, keep your least acceptable settlement position a closely guarded secret, even as you try to determine—by research (if possible) or by intelligent, imaginative guesswork (if necessary)—the least the other person will settle for. As for your maximum, the point at which you will begin the negotiation, start at as high a position as you can support—that is, persuasively justify.

The most successful negotiators are neither those willing to accept the least nor those who aim at the middle, but those who have the highest aspirations and expectations. Aim as high as you *reasonably* can. If nothing else, this will tend to give you the widest range of options and outcomes, the greatest space in which to make a good decision.

IN THE CHAPTERS THAT FOLLOW, we will look at decisions made by leaders from the realms of diplomacy, agriculture, economics, politics, war, business, and technology. Some of the decisions produced positive results, some negative, some both, some entirely unexpected. The fields covered in these chapters are diverse, and so are the decisions made within their context. Nevertheless, a common thread runs through them all. The best and boldest decisions, the decisions that produced the most impressive results, were made with a keen awareness of the matrix of factors—intellectual, emotional, political, physical, technological, and financial—in which the effects of the decision would certainly be played out. In contrast, those decisions that produced less successful outcomes reflected a more limited awareness of that matrix. Vastly unequal in result, both kinds of examples are nevertheless equally useful to those of us who want to learn as much as we can about decision making in complex, high-stakes environments: the very world in which we all live, move, and decide.

CHAPTER 2

Bismarck's Brain

The Birth of Modern Diplomacy

Previous page: *"Dropping the Pilot." Caricature of Otto von Bismarck by Sir John Tenniel, first published in the British magazine* Punch, *March 1890.*

Bismarck's Brain

THE ONLY PIECE OF ADVICE you're more likely to hear from an investment guru than "buy low, sell high" is "never get emotional about investments." It is good advice—as far as it goes. But it doesn't begin to go far enough. The secret of the great strategists, whether they play with money or with nations, is that while they avoid acting on their own emotions, they acknowledge, endeavor to become master of, and strive to manipulate the emotions of others. By all means, don't *you* get emotional about decision making. Just make sure you know how to use the emotions of *others*. As the great investors get great by using what is often called OPM—Other People's Money—so the great decision makers in any high-stakes field get great by using what we might dub OPE: Other People's Emotions. Done right, motivating other people to invest rather than relying exclusively on your own funds can bring in a lot more money. A similar principle operates when you work the decision matrix. The network of decisions becomes more powerful the farther it extends. The great decision makers—those who decide the fate of great enterprises and great nations—always make decisions that motivate, prompt, and compel the decisions of others. They are the true masters of OPE, and, of these masters, the iconic archetype was the minister-president of Prussia and the first chancellor of the German Empire, Otto Eduard Leopold, Prince of Bismarck, Duke of Lauenburg, Count of Bismarck-Schönhausen.

He was born in 1815 to substantial wealth and privilege on a Prussian estate west of Berlin, the son of a landed, retired army officer who had married a woman of even greater wealth and somewhat loftier pretensions to the lesser nobility. In the fragmented world of Germany—before he himself pulled the fragments together into a nation—Otto von Bismarck occupied an enviable yet ambiguous niche. He possessed money and a fine family name, but he fell far short of royalty. Nevertheless, he identified wholeheartedly with the aristocracy, even as he understood that he was not wholly invested in it and would not be admitted fully into its embrace. It must, therefore, have surprised his parents when he enrolled at the Georg August University of Göttingen, a hotbed of liberal—that is, anti-aristocratic—thought. We don't know why he chose this school, but we do know that he did not stay long before transferring to the more appropriately conservative Friedrich Wilhelm University in Berlin.

Young Bismarck's ambition was to parlay his education and family name into a comfortable posting in the Prussian diplomatic corps, but, as it turned out, the most he could manage was a series of jobs in government back offices in the western border city of Aachen and the Berlin suburb of Potsdam. This disappointment notwithstanding, he did follow in his father's footsteps by making an advantageous marriage, in 1847, to Johanna von Puttkamer, a noblewoman whose station was substantially higher than his own. This same year saw his first political breakthrough when he was elected to the Landtag, the newly established Prussian legislature.

A neophyte politician, Bismarck allowed himself to become emotional, embracing with singular passion the royalist cause and arguing the distinctly moribund case of the divine right of kings. But shortly after Bismarck's election, all Europe—the German states included—was swept by the wave of revolutions that marked the year 1848. Bismarck was prepared to stand fast, but the Prussian king, Friedrich Wilhelm IV, who talked tough at first, quickly knuckled under to liberal demands, not only conceding point after point to the liberals, but even agreeing to the creation of a constitutional monarchy under a unified German state.

Despite being out of step with the times—and even out of sync with the vacillating king—Bismarck won election to the Landtag again in 1849. Now, however, he was no longer a beginner. All that he had seen taught him a valuable lesson. His emotion-charged convictions clearly had not made an impact on the king. But, Bismarck reasoned, the king, having proved pliable to the liberal appeal, might well prove just as pliable to the appeal of the conservatives. Moreover, the convictions of the liberals themselves had proven largely ineffective. The revolutionary fervor of 1848, in Germany, as elsewhere in Europe, turned out to be ephemeral and quickly petered out. Snappish bickering and vicious infighting among the disjointed liberal parties brought the entire liberal movement to the verge of collapse—without the passionate Bismarck or any like-minded royalists having lifted a finger against them. Emotion—true belief, absolute conviction—Bismarck saw, was one thing, but political reality was quite another. It was far more changeable. That meant it could be manipulated and it could be played, provided that the player remained sufficiently cool, detached, observant, aware, and, at all times, intellectually ruthless.

A few years would go by before Bismarck found a word for this political reality and the art of its manipulation. In 1853, August Ludwig von Rochau, a journalist and politician, published *Principles of Realpolitik, Applied to the Political Conditions of Germany*. *Realpolitik* became the label for a strategy of identifying and pursuing attainable political, social, and economic goals not through emotion-charged idealism or ideology, not even through allegiance to one's own aristocratic family background, nor even through patriotism, but by whatever means and methods presented themselves as available at a particular time. This meant divorcing yourself from your own feelings while intimately knowing and relentlessly exploiting the emotions of others. Beginning in 1849, even before the term *Realpolitik* was coined, Bismarck became a virtuoso of this strategy.

The man who almost single-handedly instigated and orchestrated the unification of Germany was, in 1849, actually opposed to it. He argued that unification would do nothing more than compromise the independence of Prussia. Nevertheless, he lobbied for and accepted appointment to the Erfurt Parliament, the very assembly tasked with planning unification. His purpose was to understand the emerging unification proposals from the perspective of an insider, so that he could more effectively counter them. What he soon came to understand was that the threat of unification was actually slight. The Erfurt Parliament was doomed because the largest and most powerful German states, Prussia and Austria, were, like him, opposed to unification.

It was yet another lesson. Emotion was emotion, but power was power. And, when it came to making a decision and implementing it, power always trumped emotion—though (and this was supremely important) in order to come to control power, you had first to acquire control of the collective emotions of the nation.

Emotion is a coin with two heads. It can be mercurial, and therefore unreliable. Or it can be stubborn, clinging to ideas, concepts, and positions long after they cease to be useful. The result? Emotion is unreliable.

Although Germany was not unified, the German states had created a German Confederation, a loose alliance rather than anything resembling a nation in the modern sense. In 1851, King Friedrich Wilhelm IV appointed Bismarck as Prussian representative to the Confederation, which was convening in Frankfurt. He served in the Confederation well into 1858, a period during which he was, remarkably enough, entirely won over to the cause of unification. This change of heart and mind came about, in part, because he was separated from his Berlin-based reactionary brethren and therefore enjoyed greater intellectual freedom. More important, Bismarck's sojourn in Frankfurt opened up the larger picture to him. In Berlin, he was a local politician, a conservative Prussian who felt no need to look beyond Prussian borders. From the perspective of Frankfurt, however, he came to see that Prussia's survival, let alone its dominance, depended on developing a genuinely nationalistic connection with the other German states, lest Germany itself, Prussia included, be swallowed up by Austria, whose influence was growing steadily. To save Prussia, Bismarck realized that he had to look beyond Prussia. By the height of his power, he would acquire the habit of always looking far beyond it.

After King Friedrich Wilhelm was disabled by a stroke in 1857, his brother Wilhelm assumed power as regent, recalled Bismarck from Frankfurt, and appointed him ambassador to Russia. Now Bismarck's horizons widened even farther, becoming broadly international in scope. This was a momentous promotion for Bismarck, and Wilhelm, as bold as his brother had been timid, took other momentous steps. He elevated Helmuth von Moltke to the position of chief of staff of the Prussian army, and he made Albrecht von Roon minister of war. Moltke was a profound strategist and logistician, who would prove himself one of military history's great captains. As for Roon, he was a dedicated militarist, an administrator determined to shape the Prussian military into an instrument of armed diplomacy. Together with Bismarck, Moltke and Roon were the moving forces behind the creation of the world's most modern and formidable military machine. They ushered into existence the tradition of militarism that would ultimately guide Germany, fatefully, into the twentieth century. For the more immediate future, Wilhelm had put into place a triumvirate that would shape Prussian foreign policy and transform the kingdom into the nucleus of a powerful empire.

In 1861, Friedrich Wilhelm IV died, and the regent now took the throne as Wilhelm I. Whereas his brother had bent readily to the liberal will, the new king just as stoutly resisted it. For its part, the liberal-dominated Prussian Diet in 1862 defied Wilhelm's plan to reform and expand the army under Moltke. Neither side would compromise, and a standoff developed. Believing that Bismarck and only Bismarck would be able to bring the Diet into line,

Wilhelm pondered recalling him from St. Petersburg to become minister-president—effectively prime minister—of Prussia. But Bismarck had adopted an inflexibility of his own. He insisted that he would accept nothing less than a dual appointment as minister-president and foreign minister. His time in Russia had persuaded him that, to be effective, the political leader of Prussia would have to exercise control over more than the Prussian interior. Effective leadership demanded control over foreign policy as well. Taken aback by the scope of Bismarck's ambition, the new king hesitated to recall him. What finally moved Wilhelm I was the Prussian House of Deputies, which, in September 1862, overwhelmingly rejected the king's proposed military budget. Fearing for the future of the army, Minister of War Roon appealed to his king: *Bring Bismarck home.* And so it was that, on September 23, King Wilhelm I appointed Otto von Bismarck both minister-president and foreign minister of Prussia.

According to legend, Alexander the Great solved the insoluble—how to untie the incredibly complicated Gordian knot—with an unexpected but utterly effective stroke of his sword. Bismarck did much the same thing. Instead of attempting to unknot the objections and arguments of the liberal legislators, Bismarck simply refused to deal further with the Diet or the House of Deputies. Instead, he pointed to the fact that the Prussian constitution was silent on the circumstances of the present contingency—the refusal of the legislature to approve a budget—and that, therefore, he was free (indeed, he claimed, obliged) to apply the budget of the preceding year to this. And so it would go for the next four years. Each year, the legislators refused to approve the king's budget, and each year Bismarck ordered the Prussian government to collect the same taxes it had collected based on the last approved budget, that of 1861. And each year, the minister-president turned over the bulk of this money to the army to finance the creation of the military machine Moltke and Roon had envisioned.

In 1863, the House of Deputies manfully enacted a resolution declaring that it could no longer work with Otto von Bismarck. Ideologically, this should have come as a crippling, even fatal blow, but Bismarck had long ago seen beyond ideology, and, by giving Wilhelm I the budget and the army he wanted, Bismarck had taken the Prussian government in the direction of his vision, from ideology to *Realpolitik*. Yes, there was a constitution. But results, not the constitution, were the far more relevant reality. Thus, instead of bowing to the Deputies and dismissing Bismarck, the king dissolved the legislature, charging *it* with having violated the constitution by attempting to overthrow the king's minister. Armed with what he took as a royal grant of virtually limitless authority, Bismarck tightened his grip by issuing a peremptory decree sharply abridging freedom of the press. The most effective response to a public outcry, he reasoned, was not to answer it, but to suppress it.

This step failed—but, then, it also succeeded.

It failed inasmuch as the Prussian people elected to the legislature a liberal coalition ticket in 1863, and, no sooner had they taken their seats, than the newly elected lawmakers bombarded King Wilhelm I with petitions for the dismissal of Bismarck. Bending his head into the storm, the king refused to yield. He believed that it was bad enough to be saddled with a liberal legislature. He was not about to clear the way for installation of a liberal ministry as well.

The Prussian state was thus divided—never a good thing—but the answer to this division, the creation of solidarity among the Prussian people, Bismarck increasingly saw, was not to be found within Prussia, but without, beyond the borders of the kingdom. Early in his ministry, Bismarck had declared that "the great questions of the day"—paramount among them the unification of the German states into a single German Empire—would "not be decided by speeches and resolutions of majorities," the unsuccessful methods of the failed Frankfurt Parliament of 1848–49. They would, he asserted, be decided by "blood and iron." Bismarck looked for a way to employ these two elements, blood and iron, to unify Prussia under his will by building a united Germany around a Prussian core. It would be a work of diplomacy driven not by speeches and resolutions, but by military might: blood and iron.

But even Otto von Bismarck could not simply will into being a war of conquest. He needed a plausible cause for war, and so he bided his time until November 1863, when Frederick VII, king of Denmark, died, throwing into dispute succession to sovereignty over the small but important duchies of Schleswig and Holstein. Both Frederick's heir, Christian IX, and a German duke, Friedrich von Augustenburg, laid claim to rule over the duchies. Naturally, Augustenburg was the sentimental favorite of the Prussian people, but Bismarck steadfastly refused to yield to emotion. He believed that international law, a certain set of treaties, and, ultimately, Prussia's rightful place among European nations all ran counter to popular opinion. Bismarck took the remarkable step of asserting the legality of Christian's claim under an existing document, the London Protocol. Yet, even as he upheld the rule of international law, thereby protecting Prussia's standing among nations—albeit at the price of increasing his unpopularity at home—Bismarck nevertheless objected to Christian's intention of formally annexing Schleswig-Holstein to Denmark. Yes, Bismarck conceded, Christian might rule the duchy as a duke, but he was not entitled to attach it to his kingdom and rule it as a king. To reinforce his position, Bismarck turned to Prussia's chief rival for preeminence among the fractious German states, Austria, and appealed to an emotion he knew to be all-powerful in others: nationalism. Together with the Austrian government, then, the government of Prussia presented Christian IX with an ultimatum: reverse the royal annexation of Schleswig-Holstein—or face war.

Christian IX acted just as Bismarck had expected—and hoped—he would. Responding to the emotion of his own nationalism, he made the decision for war.

The Schleswig-Holstein War (also called the Danish-Prussian War) erupted early in 1864 and was over by August, Prussia (with the help of Austria) having inflicted on the Danes a defeat both stunning and humiliating. In hammering out the peace settlement, diplomats and politicians pondered putting the future of Schleswig and Holstein into the hands of the Diet of the German Confederation, thereby giving all the German states a say in the fate of the duchies. Bismarck deftly outmaneuvered the diplomats, the politicians, and the Diet as well, personally persuading the Austrian government to sign onto the Gastein Convention, by which Prussia took over administration of Schleswig and Austria, Holstein.

As an acquisition of power and territory and a demonstration of the formidable instrument that was the Prussian army under Helmuth von Moltke, this was a step in the right direction. But dividing the duchies with Austria, Prussia's chief rival, was hardly an ideal solution. Fortunately, Austria provided an occasion for a fresh exercise of *Realpolitik*. In 1866, the Austrian government suddenly had second thoughts about the status of Schleswig and Holstein and now demanded that the issue be resolved by the Diet of the German Confederation after all.

The Chinese character signifying "crisis" also signifies "opportunity." Was Otto von Bismarck aware of this fact? We do not know. But this duality describes Bismarck's approach to crisis perfectly. For him, crises were always opportunities. The conflict with Denmark had yielded quick military victory and a step toward German unification under Prussia. Bismarck now saw that a conflict with Austria would enable the next step. Therefore, charging Austria with reneging on the Gastein Convention, he ordered Prussian troops to occupy Holstein. Predictably, this created popular outrage in Austria, which appealed to the other German states for help in resisting Prussia.

A new war, the Seven Weeks' War, began on June 14, 1866, pitting Prussia (allied with Italy) against the combined forces of Austria and six other German states. Throughout Europe, the smart money was on the Austrian alliance. But the smart money had not reckoned on the army Moltke and Roon, enabled by Bismarck, had fashioned. A single overwhelming victory at the Battle of Königgrätz (sometimes called the Battle of Sadowa) crushed the forces of Austria and its allies, and by the resulting Peace of Prague, the loose and weak German Confederation was dissolved. To Prussia fell not only both Schleswig and Holstein, but also Frankfurt, Hanover, Hessen-Kassel, and Nassau. For its part, an abashed Austria was compelled to pledge future noninterference in the affairs of the rest of Germany.

Having achieved for Prussia a position of strength, Bismarck created a new coalition in 1867, the North German Confederation, under which all the German states, excluding those of the south—Austria, Bavaria, Württemberg, Baden, and the southern parts of the Grand Duchy of Hesse—coalesced around Prussia. In effect, this was a new nation, which adhered to a *Zollverein*, or customs union that removed all trade barriers between the states. Wilhelm I was president of the Confederation, with Bismarck its chancellor. Austria was left out in the cold, slipping from the position of rival of Prussia to economic dependent on the vastly more powerful Confederation. Cutting off and isolating a rival would become the trademark of Bismarck's diplomacy.

International *Realpolitik* had the effect of creating a new political reality domestically as well. Success in the form of military victory and rapidly developing industry (which had come into existence in significant measure to serve the needs of the military) brought Bismarck the popular political support he had never before enjoyed. In 1866, the House of Deputies lost its liberal majority. Among the first things Bismarck asked the obliging new conservative House to do was to vote retroactive approval of the past four years' worth of budgets.

To the victor, it is said, belong the spoils. Bismarck's all-encompassing vision took in the big picture while overlooking not the smallest detail. One of the results of the Seven Weeks' War was Prussian annexation of the Kingdom of Hanover, which had been allied with Austria. The deposed Hanoverian monarch, George V, was shown the door, but he was not expected to pass through it empty handed. Bismarck cut a deal whereby the former king retained half the assets of the crown, the rest being transferred to the coffers of the North German Confederation. Two years later, however, Bismarck charged George with scheming against the Confederation, and he summarily confiscated the remainder of the former king's assets, some 16 million thalers. Instead of depositing the confiscated funds into the general treasury of the Confederation, Bismarck appropriated them and surreptitiously deposited them into a secret slush fund he dubbed the "Reptilienfonds"—the "Reptile Fund." From this stash, personally and solely administered by himself, Bismarck drew cash for the purpose of bribing journalists to write stories favorable to his administration and unfavorable to all challengers. In 1870, when the eccentric and extravagant "Mad" King Ludwig II of Bavaria raised objections to the elevation of Wilhelm I as emperor of Germany, Bismarck drew on the fund for a bribe that instantly pacified the recalcitrant king. Bismarck retained control over the Reptile Fund throughout the duration of his administration.

The political ascendency of Prussia could hardly pass without notice all across Europe. And that was precisely what Bismarck was counting on. His ultimate object was to create stability in a European continent over which

Germany would reign supreme, but he believed the way to achieve this pacific state was by first stirring up international political instability and outright war. Of all nations, France had the most to lose in a Europe over which Germany was dominant. Bismarck understood that war with France was a growing possibility, and, although he was determined not to overplay his hand by preemptively declaring war himself, he was eager for Napoleon III to do so. To provoke this, Bismarck played on the French emperor's volatile emotions and the intense nationalist fervor pervading France. He knew that war with France—the traditional adversary of Germany—would also stir emotions in every German heart, prompting the people to rally behind Wilhelm I, who presided over the most powerful military force in Europe.

It was, Bismarck realized, a mere matter of time before a compelling reason for war would present itself. An 1868 revolution in Spain had left that nation's throne vacant. In 1870, the new Spanish government, looking for an outsider to serve as head of state, offered a German prince, Leopold of Hohenzollern-Sigmaringen, the throne. Snapping at the bait, the French government imperiously moved to block Leopold's ascension and issued a demand that the North German Confederation guarantee that no Hohenzollern would ever mount the Spanish throne. As the Confederation withheld its response, the demand became a cause célèbre throughout France, and on July 7, the French ambassador approached Wilhelm I at Bad Ems, a German resort in which the monarch was vacationing. The ambassador demanded the king's promise to withdraw the candidacy of Leopold. Wilhelm replied indignantly that he could not make such a promise, since he was not responsible for Leopold's candidacy in the first place. The ambassador returned to Bad Ems on July 11 to press his demand again. As it turned out, there really was no need for him to do so. Unwilling to provoke a crisis, Leopold had voluntarily withdrawn himself from consideration without having been asked. That should have settled the matter and defused the crisis, but the French foreign minister managed to snatch defeat from the jaws of victory by insisting that his ambassador extract from Wilhelm I a formal agreement to Leopold's demurral in addition to a pledge that the Confederation would never support a Hohenzollern as Spanish king.

It was a ridiculous request—and Bismarck made the most of it. He released to the press—on July 14, Bastille Day, no less—his personally edited version of a telegram the king had sent from Bad Ems to the Prussian foreign office reporting his confrontation with the French ambassador. In his version of the telegram, Bismarck transformed Wilhelm's refusal to assent to a formal agreement into something that approached the official severance of diplomatic relations—almost inevitably a prelude to war. Bismarck intended publication of this version of the "Ems Telegram" to function as what he called "a red rag to the Gallic bull." It did. Just five days after the telegram appeared in print, France declared war on Prussia.

The German states responded as Bismarck had imagined and hoped they would. All rallied behind Prussia, which, in the space of a five-month campaign, won every battle of the Franco-Prussian War. In surrendering, France forfeited a large cash indemnity and lost Alsace as well as part of Lorraine. Even worse, the war provided the occasion—the ideal conditions of *Realpolitik*—for the final unification of Germany. All of the states of the North German Confederation fell into line, and on January 18, 1871, Wilhelm I was proclaimed Kaiser—emperor—of the German Empire in a ceremony Bismarck insisted on holding in the vast Hall of Mirrors at Versailles. Thus the humiliation of France was complete.

In the new empire, Otto von Bismarck was created a prince and named imperial chancellor while also retaining his original Prussian offices, minister-president and foreign minister. He now worked to establish the stability of a peaceful Europe, a kind of *pax Germanica*, which would contribute to the long-term prosperity of the new empire. He well understood that France, no matter how badly humiliated it was at present, was still a force to be reckoned with, and so he sought to structure the decisions whereby the German Empire conducted its foreign policy always with the goal of isolating France diplomatically. To this end, he extended a hand of friendship to Austria and to Russia as well. To avoid alarming or alienating Great Britain, he forbore taking steps to acquire a colonial empire, lest he give the appearance that Germany meant to compete with Great Britain on the world stage. For the same reason, he chose not to expand the German navy, and thus he allowed Britain to remain supreme on the high seas. Secure in their own empire protected by their great Royal Navy, the British did not move against the bold expansion and consolidation of German power.

The new stability Bismarck had created lasted for a little while. But when Russian victories over the Ottoman

Empire in the Russo-Turkish War of 1877–78 threatened to give Russia too much power and influence, Bismarck cooperated with other European statesmen to check the ambition of that nation. As a result, Russia drifted away from the German orbit and into that of France. Moreover, by the end of the 1870s, Bismarck found that he could no longer resist the tide of German popular opinion, which continually clamored for the creation of a colonial empire. Germany accordingly reached out into those parts of Africa as yet unclaimed by other European powers. By the time Wilhelm I died in 1888, the *pax Germanica* of Otto von Bismarck was beginning to show signs of giving way. Friedrich III, Wilhelm's son, succeeded his father, but within three months succumbed to cancer, and his place was taken by his son, Wilhelm II.

The second Wilhelm was a very different man from the first. Whereas the earlier monarch had been patient in allowing Bismarck to create and exploit the circumstances of *Realpolitik* that were congenial to the careful creation and maintenance of an empire, Wilhelm II was eager to rush ahead in a program of aggressive expansion that took little or no account of what the rest of the world thought. This put a great strain on relations between the new emperor and his prime minister, but the final rupture between Bismarck and Wilhelm II came not over foreign policy but as a result of domestic matters. Tellingly, the final breach opened up as soon as Bismarck violated his own rule about keeping his emotion separate from his vision. Wilhelm II sought to make certain accommodations with liberal and labor elements in Germany. Warning that doing this would pave the way for the rise in Germany of the socialist movement that was emerging elsewhere throughout Europe, Bismarck, in defiance of the Kaiser, set about the enactment of extensive and severe anti-Socialist laws. He managed to persuade a legislative majority to sign onto the legislation, but a bitter split developed over whether or not the police should be authorized to evict socialist agitators from their own homes. Bismarck insisted that this authority was absolutely necessary, and dug in, refusing to make any compromise on this provision. In contrast, Wilhelm II announced his willingness to yield on this highly unpopular point. The emperor secured passage of the legislation–minus the police authority of eviction. Bismarck would not give up. He now pushed and pushed the emperor to veto the compromise bill. The Kaiser protested that he could not understand why he should do this. Even without the eviction provision, the legislation provided unprecedented police powers against the socialists. An argument developed, tempers grew short, and, at last, Bismarck blurted out to the Kaiser: He wanted the bill to fail so that the socialists would play their hand by staging a riot or some other violent demonstration, which would serve as ample reason simply to crush them by force in a single, overwhelmingly harsh blow.

Expressed in the heat of emotion, this naked motive of violent repression, coupled with Bismarck's easy willingness to practice subterfuge of the most cynical sort, repulsed even the impatient new Kaiser. Responding angrily that he was not about to commence his reign with an orgy of bloodshed, he signed the compromise law in 1889 and, what is more, the very next year compelled Otto von Bismarck to resign all of his offices.

Not since Napoleon I had the politics of Europe so entirely reflected the influence of a single figure. The departure of Bismarck sent shockwaves across the continent. A British newspaper printed an instantly famous cartoon that showed Otto von Bismarck descending the gangway of the German ship of state. "Dropping the Pilot," the caption ran.

FOR YEARS, THE DECISIONS of Otto von Bismarck had largely determined the fate of Europe. His exceptionally influential and effective decision making had depended on an acute understanding of local conditions but always within a global context. It was a process that required reducing continental politics, including the bloody business of war, to a kind of game in which it was necessary to take an intense intellectual interest—the most exacting and unbroken concentration was an absolute requirement—yet an interest utterly divorced from one's own emotions. For these only tended to blur the global aspects of vision. In the end, even Bismarck was unable to maintain both the extreme degree of concentration and the absolute emotional detachment his strategic methodology demanded. Moreover, his goal of creating a stable Europe proved elusive. The needs of nations were too dynamic, and the emotions of their leaders too unpredictable. If anything, the network of alliances and enmities his long diplomacy had created would serve to propel the continent and most of the world into the cataclysm of the Great War of 1914-18 as well as a second world-engulfing war some twenty years after the first. Yet there is much to be learned from the example of Bismarck, who always maintained himself ready and able to use whatever the world offered–whether in the form of aid, obstacle, or both–as he made the decisions that led his nation toward whatever goals he envisioned and defined.

TAKEAWAY

BISMARCK WAS ONE OF HISTORY'S great visualizers. Like the adept player of RISK, he looked at the map of Europe as it was and envisioned that map as he wanted it to be, then he played the "great game" of diplomacy to create that redrawn vision. In this, he played on the collective emotions of other nations to prompt, cajole, or coerce the other powers of Europe to make decisions that ultimately led to the elevation of Germany as the greatest political and economic force on the continent.

CHAPTER 3

The Story of Kudzu

All About the Plant That Ate the South

Previous page: *Bird's-eye view of Fairmount Park, Philadelphia, at the Centennial Exhibition, 1876.*

The Story of Kudzu

IT CERTAINLY SEEMED LIKE A NO-BRAINER at that particular time and in that particular place. In 1876, the United States was celebrating the centennial of its independence by throwing a world-class cultural and commercial blowout: the Centennial Exhibition, America's first major world's fair, in which the nations of the globe were invited to Philadelphia to honor the nation's birthday by showing off the best *they* had to offer. Japan built a lovely garden in Philadelphia's premier green space, Fairmount Park. It was a big hit, and the garden plant that made the greatest impression was something the Japanese called *kuzu*, after the area, near modern Yoshino (Nara prefecture), in which the plant was first known and cultivated hundreds of years ago. To American ears, the name sounded like kudzu, and, in the spirit of cultural imperialism that characterized the age, kudzu it became.

It is certainly an attractive plant. A climbing vine, it spreads up just about anything vertical and across just about anything horizontal. It produces very pretty, long, purple flowers that are rich in nectar and refulgent in fragrance. America's gardeners loved it as soon as they saw it in 1876 and began planting it everywhere. But it was in the 1920s that Charles and Lillie Pleas, who operated the Glen Arden Nursery in Chipley, Florida, envisioned a truly vast market for the plant. They discovered that kudzu was not merely decorative, it was nutritious—at least animals liked to eat its leaves. So the enterprising nursery owners developed and advertised kudzu as a new, fast-growing, very inexpensive, and nearly indestructible forage feed. They offered kudzu for sale through catalogues and promoted it with special mailings nationwide, but especially in the South.

The fact is that kudzu has many more uses than either as decoration or forage. Indeed, in Asia, it has been cultivated for so many purposes that, on paper at least, it seems to be a miracle plant. Kudzu is edible by animals as well as human beings. Its leaves are suitable for salads, kudzu flowers can be dipped in batter and fried, and starchy kudzu roots can be cooked and eaten like any tuberous vegetable. Kudzu blossoms can also be processed into a thick, golden jelly with a unique flavor that connoisseurs compare to that of bubblegum. In Japan and China, the roots of kudzu have long been ground into a powder that is used in any number of herbal medicines. Modern pharmacologists, both in the East and West, have found that consuming kudzu tends to reduce the craving for alcohol and also is one of few substances that is truly effective in relieving a hangover—something that is necessary, presumably, if one fails to eat one's kudzu *before* embarking on a bender. In any case, as of winter 2007, Harvard University researchers were investigating kudzu derivatives as an adjunctive therapy in treating chronic alcoholism.

And there is yet more. The anti-inflammatory and antimicrobial isoflavone daidzein may be extracted from the plant, as can another isoflavone, the powerful antioxidant daidzin, which is widely believed to help prevent cancer. Another isoflavone, genistein, also available in kudzu, has been shown to be effective in treating leukemia. Other kudzu derivatives are used to treat headaches, including migraines, and also tinnitus (ringing in the ears) and vertigo. In medieval Japan, the plant was considered so valuable that it was routinely harvested and collected as a tax.

Although the Pleas of Florida had initiated the first extensive spread of kudzu cultivation in the United States, and while others discovered a host of additional uses for the plant, it was not ultimately as an ornamental garden plant, or animal feed, human food, or even as a medicine that kudzu came to explode into the American South. The final triumph of kudzu was largely the result of a concerted effort by the U.S. government.

From 1933 to 1939, in the depths of the Great Depression, America's farm lands were raked by the great dust storms of the so-called Dust Bowl. A combination of severe and prolonged drought and years of the farming of such nutrient-intensive crops as cotton, corn, and grain without prudent crop rotation severely depleted millions of acres of topsoil, turning it to dust, which blew away. Desperate to identify emergency means of preventing further erosion and reclaiming the soil, the U.S. Soil Conservation Service eagerly seized on kudzu. It looked to be another no-brainer. Not only did the rapidly growing kudzu vines, their long creepers shooting out horizontally in all directions, tie together drought-starved topsoil that would otherwise blow away, the plant acted symbiotically with nitrogen-fixing bacteria already present in soil to increase the available nitrogen content and thereby render the soil more fertile. In addition, the roots of the plant worked to bring up minerals from below the surface and into the topsoil, further rein forcing the topsoil structurally while also enhancing its nutrient content. Kudzu appeared to be not only an erosion-stopping machine, but a soil reclamation machine as well.

The Civilian Conservation Corps (CCC)—the federally funded army of young men employed nationally during the Depression in vast programs of public works—planted kudzu far and wide, and, well into the 1940s, the government paid individual farmers the spectacular price of eight dollars an acre to plant fields of kudzu.

Kudzu, this Japanese import, rapidly became an article of American patriotism. Not only did the U.S. government propagate and promote it, the plant also acquired a supremely eloquent champion in the person of Channing Cope. During the late 1930s and throughout the 1940s, as farm editor of the *Atlanta Constitution*, the portly Cope—as of 1949, he weighed in at 205 pounds—was one of the most influential newspapermen in the South. His farm column earned him an early-morning spot on WSB-AM radio, and he rose daily at five in the morning, donned a trademark suit of rumpled seersucker, descended from the bedroom to the kitchen of his Covington, Georgia, farmhouse, poured himself a cup of coffee laced liberally with bourbon, drank it down, then started reading aloud from his newspaper in order to clear his pipes for the broadcast that commenced at precisely 6:01. Weather permitting, the show originated from an easy chair on Cope's screened front porch. After eight bars of "Dixie" and a brief commercial announcement, Cope would begin, typically with something like "I never saw a prettier day . . ." The broadcast over at 6:30, Cope banged out his daily column, "Channing Cope's

Almanac" for the *Constitution*, which syndicated it nationwide. The homespun philosopher of farming exercised even more influence through a series of books, the best known of which was *Front Porch Farmer*, which not only became a popular bestseller, but was adopted by at least four southern state universities as an agricultural textbook.

If Cope preached the religion of farming, his favorite gospel was the advent of a reinvigorated southern agriculture made possible by rich new topsoil, year-round pasturage, and an abundance of milk-fed beef. All of this, Cope claimed, would be founded on the "miracle vine," kudzu, about which Cope was so enthusiastic that he founded the Georgia Kudzu Club, which boasted, at its height, more than twenty thousand members. With the zeal of a preacher and the tenacity of a press agent, he hawked the vine so relentlessly that his friends, listeners, and readers dubbed him the "Kudzu Kid." Both respected and colorful–his columns blended authoritative and well-informed farming advice with disquisitions on barbershop quartets and alcoholism, as well as a memorably eloquent epitaph on his deceased Scottie, Mr. Burns–Channing Cope was an extraordinarily effective propagandist, doing for kudzu in the American southland of the twentieth century what Johnnie Appleseed had done for the humble apple tree in the Midwest of the early nineteenth century. When Cope wasn't writing about kudzu or broadcasting about kudzu, he stumped the country, starting Kudzu Club branches nationwide to promote what he believed was the new principal southern crop. "Cotton isn't king in the South anymore," he proclaimed. "Kudzu is king!"

The trouble is a king can become a tyrant.

The decision to introduce and to promote the spread of kudzu began with an appreciation of the plant's beauty, then proceeded to an ever-widening awareness of its utility, culminating in near worship, as the vine began to appear as a panacea, especially in the areas of livestock feed and soil-erosion control and enrichment. For the likes of Channing Cope, the plant seemed to contain no less than the seeds of his region's very future. We have all been told, over and over, that you really can get too much of a good thing, but believing this when it comes to making decisions seems to be another matter altogether. By focusing on the beauty and utility of kudzu, the plant's early promoters, ranging from government scientists to homespun print and broadcast celebrities, overlooked the bigger picture: the effect of environment and the effect on environment. Kudzu grows fast. In the American South, which offers a perfect climate for the plant—hot, humid summers, ample rainfall, mild winters, few hard freezes—kudzu grows with truly astounding velocity. A foot per twenty-four hours is not uncommon. Moreover, while it can serve as forage, in the South, kudzu has no natural predators. There is nothing to stop it. This means that kudzu, artificially introduced into an environment to which it was not native, took hold as what ecologists call an invasive species.

This is not always a bad thing. Corn, wheat, rice, and a host of other familiar food crops, as well as cattle, poultry, and some other species of livestock were all introduced to North America by settlers. None of these species is native. Today, they account for more than 98 percent of the nation's food system. Yet it is also true that, worldwide, approximately 80 percent of those species classified as endangered would suffer catastrophic losses if they had to compete with—or were preyed upon by—invasive species. Introducing species from one environment to another is always risky, and it is very difficult—sometimes impossible—to envision the impact on the new ecosystem. In terms of decision making, few people possess the knowledge, the competence, or even the basic data to gain a picture of the entire matrix in which the decision to introduce the species will develop and ramify.

The case of kudzu is hardly unique. In Hawaii, for example, a type of wetland plant called `ae`ae (*Bacopa monnieri*), though native to the islands, has become an invasive pest species in artificially managed and manipulated water bird refuges. The problem is that it rapidly overgrows the shallow mudflats that have been artificially established for an endangered Hawaiian bird called the stilt, rendering the flats undesirable as feeding areas for the birds. The decision matrix becomes even more complex and obscure when more than one nonnative species is introduced to an environment in succession. Effects can be highly interactive in unanticipated ways. For example, introducing a second nonnative species may create conditions that allow the other nonnative species, introduced earlier, to increase explosively, becoming harmfully invasive. This happened in the nineteenth century, when the amethyst gem clam (*Gemma gemma*) was brought from the East Coast to California's Bodega Harbor. For more than a hundred years after its introduction, the gem clam coexisted harmlessly with native clam species in the harbor, but in the mid-1990s, after the European green crab (*Carcinus maenas*) was introduced, an ecological crisis was created. The green crab preyed primarily on species native to Bodega Harbor, resulting both in a dramatic decline in the native clams as well as an explosive increase both in the green crab *and* the earlier-introduced amethyst gem clam.

Unforeseen ecosystem impacts can, of course, create profound economic consequences, including lost production in agriculture and forestry as well as the costs of measures to attempt management, control, or eradication of the invasive species. It is estimated that managing invasive species costs nearly $140 billion in the United States alone. The total cost of lost production is impossible to calculate, but is certainly far greater. Indeed, in addition to lost agricultural and forestry costs, invasive species often have a devastating effect on recreational activities and industries, including tourism, fishing, hunting, hiking, and waterborne recreation.

Most immediately alarming are the potential health effects of the proliferation of certain invasive species. In some cases, it is human beings who are the encroaching offenders. For example, epidemiologists note that the spread of the AIDS virus into human populations may be traced directly to increased access to previously remote regions of Africa. It is also the case that exotic birds, insects, and rodents introduced into a new environment—intentionally imported or carried unintentionally in a ship's cargo, for instance—have been vectors in catastrophic epidemics and pandemics of such scourges as yellow fever, typhus, malaria, and, in the Middle Ages of Europe, bubonic plague—the Black Death. More recently, the West Nile virus has come to North America and spread across the continent, vectored by mosquitoes, and causing numerous human fatalities in addition to wiping out various bird, mammal, and reptile populations.

On a worldwide scale, the action of invasive species is responsible for significant reductions in biodiversity. In an era of globalization, with the transportation of goods as well as people over vast distances extremely common, the impact of invasive species on biodiversity has steadily increased. Ships traveling from port to port often empty their ballast water indiscriminately, sometimes introducing a host of potentially invasive species into a particular harbor.

Incidental and accidental introduction of nonnative species is one thing, but purposely importing a species, as in the case of kudzu, can have even more dire consequences. The problem is that, in most cases, those who introduce the new species focus on the benefits of the particular plant or animal rather than attempting to factor in the environmental context. As mentioned, predicting long-term consequences is often difficult or even impossible, especially since there is no telling what other new species may someday be introduced, which—as in the case of the clams—can radically alter the ecosystem. The species that is introduced, after all, has no knowledge of the motives of those who have introduced it.

In some environments, especially where a lot of the native species have evolved highly specialized requirements, nonnative plants and animals rarely become invasive. Unable to compete with the native species, they soon die out. But the ecosystem of the American South does not present a high degree of specialization–many species, both native and imported, compete for the same resources–and for this reason, kudzu, finding a congenial environment, exploded, often displacing native plants. One attempt after another to curb or eradicate the plant failed and continues to fail. Since 1953, when the U.S. Department of Agriculture reclassified kudzu from miracle plant to "pest weed," experiments with traditional herbicides have been conducted and found to be, in a word, disappointing. Most have virtually no effect, and at least one common chemical tested actually promoted growth. Mowing can control the spread of kudzu, but it must be thorough and weekly, and carried out on a regular basis for years before it has anything like a permanent effect. Some southern cities have tried bringing in grazing animals. Chattanooga has purchased goats and llamas for this purpose. Both species like to graze on the plant, and the llamas protect the goats against their natural predators. Recently, a common fungus, *Myrothecium verrucaria*, has been found to be effective in killing kudzu, but the fungus also kills cotton, and it will require careful environmental and toxicological studies to determine not only whether it may be safely introduced, but just how and when to do it.

In the meantime, the kudzu keeps growing. It continues to overgrow much of the Southeast, reaching as far north as New Jersey and Pennsylvania, as far south as the Gulf Coast, as far east as the Atlantic seaboard, and as far west as eastern Oklahoma and Texas, nearly twenty thousand square miles in the United States. The plant bears seeds, but it spreads principally by runners (stolons) and rootstalks (rhizomes), which grow at a prodigious rate.

ALTHOUGH THE PROLIFERATION OF KUDZU IS DRAMATIC—it is an outstanding feature of the southern landscape—as ecological problems go, the explosion of this plant is not a true cataclysm. To be sure, it is a very costly nuisance, responsible for about half a billion dollars in cropland loss and control costs annually, but most southerners have a sneaking affection for the plant, even when they curse it. Nevertheless, its spread, its persistence, and its costs remind us of the consequences of failure to consider the entire decision matrix before taking what may seem the most innocuous of steps: the importation and propagation of a very pretty, highly useful, and extraordinarily fast-growing Japanese plant. Hey, what could go wrong?

TAKEAWAY

ECOSYSTEMS AND THE ENTIRE natural environment present a vast and vastly complex matrix in which human beings make many decisions. Often, the nature of the environmental matrix magnifies the effect of a decision, which has ramifications throughout an ecosystem, sometimes for the better, but most often for the worse. Focusing narrowly on a particular action—the introduction of a miracle plant, for example—a decision is made in isolation, without pondering the great matrix of relationships that will be created, destroyed, and otherwise impacted. Typically, the response to the environmental mess that follows is to treat the problem as an emergency, as if it were an explosion or an earthquake. The trouble with this approach is that, unlike explosions and earthquakes, which are genuinely emergent—that is, virtually instantaneous—ecological catastrophes occur over time and must be remedied the same way, over time, using methods and means that do what the original decision failed to do: take into account a vast web of relationships existing in space as well as time. The bottom line? Kudzu is a great plant, but a bad decision.

CHAPTER 4

Fireside Chat

Global Strategy for the American Masses

Previous page: *Franklin Delano Roosevelt delivers a fireside chat in 1938.*

Fireside Chat

IF, LIKE MANY POETS AND PHILOSOPHERS, we picture time as a flowing river, we may also see the events of history as so many streams and tributaries, some coming together here, others diverging there, and still others pursuing their own courses and wandering in diverse directions. Looking at the panorama of history, it sometimes seems apparent that certain streams were virtually destined to merge, whereas others, widely separated from one another, clearly appear unlikely to ever be capable of joining or even of entering the mainstream. Such were, apparently, the courses of events associated with the lives of Guglielmo Marconi and Franklin Delano Roosevelt. To all appearances, their lives and works were utterly unrelated to one another, had no reason to be related to one another, and never could be related to one another.

Marconi was born near Bologna, Italy, to a prosperous Italian landowner and his Irish wife, Annie Jameson, whose grandfather had started the celebrated Jameson Whiskey distillery. Young Marconi struggled through most of his school subjects but exhibited an early aptitude for science, especially the field of electricity. His family's money allowed him to study for a time with Augusto Righi, a physics professor at the University of Bologna who had worked with no less a figure than Henrich Hertz, the great German physicist who had both propagated and detected electromagnetic radiation, and for whom the standard unit of frequency–the hertz–is named. Inspired by Righi, Marconi began pursuing the mysteries of electrical transmission without wires. More inventor than theoretician, he set about building a variety of transmitters and receivers before he made his first practical demonstration of radio in 1895, sending and receiving signals over approximately one mile. Finding neither interest in nor financial support for his work, he decided to develop instruments of greater and greater range, finally staging a demonstration that spanned the Atlantic on December 12, 1901. With this highly publicized feat, radio took off, and Guglielmo Marconi was on his way to becoming a very wealthy man.

He was hailed as a visionary, receiving the Nobel Prize in Physics in 1909, though strangely enough, Marconi failed to imagine fully the ultimate impact of his invention on culture and civilization. From the beginning, Marconi thought of radio as nothing more than a wireless telegraph. It was a means of sending messages from point A to point B, which might be thousands of miles away, somewhere inaccessible by wires, or perhaps a ship at sea. This view of the technology was, to be sure, momentous, but it fell to others to develop his invention as a medium of mass communication rather than station-to-station signaling. Marconi himself never imagined that radio could become a broadcast medium capable of providing information and entertainment to the world.

On Christmas Eve, 1906, the Canadian-American engineer Reginald Fessenden, who had invented the high-frequency alternator—the first practical voice-capable radio transmitter—broadcast a program that included music and voice. By 1910, an American named Lee De Forest, who had invented the radio tube—the heart of radio technology and all electronic applications before the invention of the transistor—was experimentally broadcasting opera performances from New York's Metropolitan Opera. Working from these foundations, amateur radio enthusiasts across America began almost routinely transmitting music, speech, and coded communications before the United States entered World War I in April 1917. By that time, the potential of radio seemed so great that President Woodrow Wilson ordered a halt to most American radio broadcasting out of fear that U.S.-based German spies would use the new technology to transmit vital information to their home country. The armistice ended the war on November 11, 1918. Two years later, on November 2, 1920, radio station KDKA in Pittsburgh broadcast the results of the presidential election. Historians generally consider this the first major commercial broadcast in the United States. With amazing speed, it ushered in a new industry and a new system of entertainment and communication that would reshape the country's politics, economy, and culture.

During the "Roaring Twenties," nothing roared louder than radio. It exploded onto the American scene as the supreme medium of popular entertainment, dishing up mostly lowbrow stuff to the masses, who eagerly consumed it. The fare of radio's early heyday was a world apart from the milieu into which Franklin Delano Roosevelt had been born, on January 30, 1882: a patrician family ensconced like latter-day feudal lords and ladies in an estate at Hyde Park in Dutchess County, New York. His mother was a loving if often suffocating presence, whose ministrations enhanced the aura of august exclusivity around the family. Through the age of fourteen, the boy was schooled at home by governesses and tutors before being packed off to

Groton, a small, elite prep school in Massachusetts, which he attended from 1896 to 1900. Like the rest of the Roosevelt environment, Groton was steeped in a tradition created by old American money. However, its founder and leader, headmaster Endicott Peabody, imparted to his young men not a sense of the perquisites of their privileged status but an ethic of public service. His message to them was that wealth created for each of them an outstanding debt to society, an obligation on the part of those who enjoyed riches to exert themselves in the creation of a better world. Franklin Roosevelt heard the headmaster's lesson and heeded it.

He went on from Groton to Harvard, graduating in the middle of the Class of 1904, having, in the course of his studies, fallen in love with Anna Eleanor Roosevelt, his fifth cousin once removed and the niece of Theodore Roosevelt. On March 17, 1905, they were married. To all outward appearances, their marriage was little different from any other union in their social circle—not exactly blissful nor visibly passionate, but certainly decorous and apparently quite satisfactory. Reality differed from appearances, however. Franklin Roosevelt's attentions soon grew divided, first between Eleanor and his perpetually overbearing mother, and later between Eleanor and other women, including even Mrs. Roosevelt's own private secretary.

Yet the union survived—and it did much more than survive. Despite its many discontents, the Roosevelt relationship was profound and, at its deepest level, very different from conventional upper-crust marriages. Husband and wife functioned as a powerful team, emotionally, intellectually, and politically. Eleanor nurtured the best that was within her husband, while she herself overcame an innate shyness–she was unusually tall, bucktoothed, and possessed a shrill voice seemingly incapable of modulation–to become a leading social activist. It was her highly developed and supremely practical social conscience that guided the development of her husband's social ethics, fanning into flame the commitment to service that had been kindled as far back as his Groton days.

Franklin Roosevelt attended Columbia University Law School until the spring of 1907, but decided not to pursue his degree after he passed the New York State bar examination. Instead, he went directly into a leading Wall Street firm. It was the expected, entirely conventional move for a young man of his social station. Yet FDR soon discovered that corporate law, although highly profitable, was, for him, unfulfilling. He found himself eagerly embracing the cases that no one else in his white-shoe firm wanted: low-paying matters for "small" clients involving claims in municipal courts, which he later claimed (according to biographer Frank Freidel) taught him first-hand about the problems of poverty. In this way, the isolated stream of his privileged life began to converge with an American mainstream that included people of all sorts, many of them poor and struggling.

His social awareness heightened, Franklin Roosevelt ran for the New York State Senate in 1910 and instantly discovered something else about himself: he was a political animal. A tireless and charismatic campaigner, he won election by a wide majority, even though he was a Democrat running in a rock-ribbed Republican district.

As a state legislator, FDR earned considerable notice as a straight-talking Progressive reformer, and in 1913 he entered into the national arena when appointed to a subcabinet post in the new administration of Woodrow Wilson. As assistant secretary of the navy, Franklin Roosevelt soon outshone his boss, the affable but aging Secretary of the Navy Josephus Daniels. Roosevelt exhibited a prodigious talent for dealing with people on all levels and of every class, from admirals and Navy Department bureaucrats to the rank and file of shipyard labor unions. During World War I, he personally toured the American front lines and pushed vigorously for enlarging and modernizing the navy—something even Daniels did not do. Roosevelt was soon singled out as a young man who got things done, and when he was tapped in 1920 as the Democratic vice-presidential candidate, running mate to Ohio governor James M. Cox, he accepted the nomination with undiluted enthusiasm. Ever the realist, Roosevelt was perfectly aware that in the isolationist political climate of America that followed the "Great War"—in some quarters derided as the "European War" or even "Mr. Wilson's War"—he and Cox stood virtually no chance of victory against Republican Warren G. Harding and his promise to return the nation to "normalcy." It hardly mattered. Roosevelt campaigned with irrepressible gusto.

Defeat was inevitable, but Roosevelt had no regrets. He had gained valuable public exposure, and after the election he returned to the practice of law, establishing his own firm and biding his time until the next political opportunity arose.

But then, without warning, the stream of Franklin Roosevelt's life took a turn he could never have anticipated. In August 1921, while vacationing at the Roosevelt family compound in Campobello, Canada, FDR contracted what his physicians believe was polio (a recent review of his symptoms suggest that he actually had another neurologically devastating disease, Guillain-Barré Syndrome).

Seriously ill, he fought for his life and won, but lost the use of his legs. His mother, meaning well, counseled her stricken son to return with her to Hyde Park, where he could live out his life in the genteel retirement suited to an invalid. Eleanor Roosevelt resisted, however, prevailing over her formidable mother-in-law and helping her husband to vanquish despair as well as overcome the defeatism of his mother and other family members.

In the end, Roosevelt not only overcame his disability, but also discovered that his affliction had served to mature him. The struggle to come to terms with his personal loss and triumph over it informed the course of the rest of his public life, giving him deeper insight into the suffering and struggle of others—an insight he could never have gained in the splendid isolation of Hyde Park, and that doubtless helped him as he led the nation through depression and war.

In 1924, Roosevelt delivered an eloquent and provocative speech, nominating Al Smith as the Democratic candidate for president. Four years later, FDR successfully ran for governor of New York, scoring a victory that was the more remarkable for having been achieved in an otherwise overwhelmingly Republican election year. In all, he was elected to two terms as governor, battling a majority Republican legislature in both terms to push through many Progressive measures, including a program of reforestation, state-supported old-age pensions and unemployment insurance, labor legislation, and the public development of electric power. With the deepening of the

Great Depression in 1931 during his second term, Roosevelt became the first governor in the nation to create an effective state relief administration. He put social worker Harry Hopkins in charge of the agency. Later, Hopkins would join longtime FDR political aide Louis Howe as one of President Roosevelt's closest confidants and advisors.

The course of Franklin D. Roosevelt's life had been a widening of his perspective on the matrix of necessary decisions that running any great enterprise involves, whether the organization is a business, a state government, or an entire nation. As governor, that decision matrix came to encompass a still-new technology, radio. It existed, it was popular—practically every family and business had a radio set, or access to one, at least. Would the governor find a use for it?

Back at the turn of the century, FDR's presidential relative, Theodore Roosevelt, had transformed the veranda of his stately home in Oyster Bay, New York, as a platform from which he delivered many a memorable yet informal speech. FDR liked the informality, but he believed that the time had passed when a leader could expect the people to come to him, and he saw in radio the means by which he could bring his message to New Yorkers, into their homes and places of business—wherever people clustered around a radio set. One of FDR's speechwriters, Judge Samuel Rosenman, recalled that Governor Roosevelt first used the radio to make an informal talk to the people of his state in 1929. Direct communication with the public and an ability to create common cause rapidly became Roosevelt hallmarks, and the radio talk would be a mainstay of the Roosevelt governorship and, later, his presidency, becoming instrumental in his leadership of the nation through the multiple crises of economic depression and world war.

During his years as governor, the radio talks had no special name, but after the first one FDR delivered as president on March 12, 1933, CBS radio executive Harry C. Butcher came up with something to call the second talk, which was scheduled for broadcast on May 7, 1933. At his urging, CBS announced it as a "Fireside Chat," and the phrase stuck. President Roosevelt had found a medium, a rhetorical approach, and even a time slot (the chats were typically scheduled for broadcast after dinner) that invited American families to gather in their living rooms and parlors, before the quasi-mythical hearthside that represented home and all they held dear, to listen to their president speak directly to them about matters affecting the entire nation and the whole world, about matters affecting *them*.

That first presidential chat was typical. It focused on an issue of intense and immediate concern to Americans: the banking crisis. Financial panic had caused a nationwide epidemic of runs on banks as worried depositors rushed to withdraw their savings before their bank followed others in failure. Broadcasting not from some great hall but from the Diplomatic Reception Room of the White House—the president's own home—FDR began, "I want to talk for a few minutes with the people of the United States about banking . . ." It would be the first of some thirty Fireside Chats he would broadcast during his presidency. He saw the form and the forum as a powerful way to explain his economic policies and political programs directly to the people, one on one, without a reporter's filtering sensibility intervening. His tactic was almost always to speak on a single issue of pressing importance for an average of half an hour, in the quiet of the evening, when most people were at home, together as a family, and when audiences were at their biggest—thirty to forty million at a time when the U.S. population numbered about 130 million. The tone was spontaneous and conversational, but in actuality, this was a carefully crafted effect that the president and his staff meticulously worked into painstakingly prepared scripts. They were so beautifully scripted that they never sounded like what they were—thoroughly contrived—but it is also a fact that the buoyant optimism driving even the most serious chats came naturally to FDR.

That first chat on the banking crisis was aimed, first and foremost, at taking the mystery and therefore the terror out of a frightening situation. FDR understood the point of view of the man and woman on the street. For most working people, he knew, the local bank was nothing more than a safe place to keep money. The widespread belief was that one's money was securely tucked away in a vault and could be had on demand. FDR explained that "the bank does not put the money into a safe deposit vault. It invests your money in many different forms of credit . . . A comparatively small part of the money you put into the bank is kept in currency—an amount which in normal times is wholly sufficient to cover the cash needs of the average citizen." February and early March 1933, however, were not "normal times," and FDR continued: "there was a general rush by a large portion of our population to turn bank deposits into currency or gold—a rush so great that the soundest banks could not get enough currency to meet the demand." Why not? "The reason for this was that on the spur of the moment it was, of course, impossible to sell perfectly sound assets of a bank and convert them into cash except at panic prices far below their real value. By the afternoon of March 3, a week ago last Friday, scarcely a bank in the country was open to do business."

FDR responded to the banking crisis by calling Congress into special session on March 5 to declare a bank holiday, begun the next day, which temporarily closed the nation's financial institutions to give them some breathing space and enable bank examiners to separate solvent banks from insolvent ones. This was followed on March 9 by passage of the Emergency Banking Act, which provided the full legal authority for the action. FDR's first presidential chat explained the purpose of the bank holiday and detailed the schedule under which sound banks would reopen the following week. The president's explanation was straightforward and unblinkingly honest. FDR believed that if the people could be made to understand the banking crisis and, just as important, what was being done about it, they would heed his plea to remain calm.

The remarkable thing was that Roosevelt was right. Not only did most of the nation's solvent banks reopen according to the schedule he outlined (some institutions that required more extensive inspection opened later), but within a few days, deposits were outstripping withdrawals. The banking crisis had been successfully managed through a combination of emergency legislation and honest, clear communication–a leader speaking to the people as one citizen to another.

FDR was a strong executive who pushed through Congress a mountain of legislation and freely issued a long series of executive orders. Nevertheless, early on he made the decision that, while he would be a strong president, he would not be a dictator. Americans would accept a strong executive, but they would flatly reject a tyrant. The Fireside Chat used the most popular of popular media—exuberantly lowbrow and resolutely democratic—to inform the nation and shape its course, decision by decision, across an often bewildering matrix of decisions that could not be dodged. His Fireside Chat on the new National Recovery Administration (NRA), broadcast on July 24, 1933, was aimed at persuading the people to overcome their fear of this radically new agency, and the approach the president took was to render the disturbingly unfamiliar in terms of the comfortably familiar.

FDR began his talk with the most basic of concepts: "this simple principle of everybody doing things together." If the new and the unfamiliar were frightening, the related ideas of help, mutual aid, teamwork, cooperation, and pooled strength were very familiar and even highly comforting. By "doing things together," FDR explained, "we are starting out on this nationwide attack on unemployment." Thus, in the informal context of a Fireside Chat, broadcast on the soothing medium of radio, FDR introduced the simple and comforting principle of cooperation in a war against the *real* enemy, the thing everyone feared and had every reason to fear: unemployment.

Having taken the first big step toward dispelling the fog of fear from the idea of the National Recovery Administration, FDR continued by emphasizing the critical importance of everyone understanding what the NRA was all about: "It will succeed if our people understand it—in the big industries, in the little shops, in the great cities, and in the small villages."

Each Fireside Chat was aimed, like this one, at getting every American fully engaged. No leader can do all the pulling by himself. Everyone must pull together. Moreover, everyone must do his or her part, beginning with an effort, at all levels, to understand. The NRA program was not just for big industries or large urban areas, but for everyone at every level. Fortunately, the president assured his audience, "There is nothing complicated about it," nor even anything "particularly new about the principle." He explained: "It goes back to the basic idea of society and of the nation itself that people acting in a group can accomplish things which no individual acting alone could even hope to bring about."

You cannot get more basic, more elementary than this. FDR appreciated that many of his political opponents saw the NRA as a radical departure from traditional American capitalism—it was, after all, a means of bringing labor, management, and rival firms into close cooperation rather than setting them at each other's throats in capitalist competition—but he presented it to the American people as an expression of a timeless and very simple idea, the idea of society and of nationhood, and the very motive that induces people to live and work together rather than separately: "that people acting in a group can accomplish things which no individual acting alone could even hope to bring about." Who could object to this, let alone be afraid of it?

THIS COMBINATION OF STRAIGHT TALK, COMPASSION, and believable optimism drove the powerful rhetorical strategy behind all of the Fireside Chats, those dealing with the crises of economic depression and, later, those dealing with war. Each outlined a problem as well as a course of action—a set of decisions to be made—and each was designed to impart the feeling of confidence necessary to make those decisions and to act upon them. Thus, the Fireside Chat provided the emotional and intellectual means not only to survive the crisis, but to triumph over it, transforming crisis into an opportunity to build a nation of greater unity and strength, with the collective ability to continue making the right decisions and the collective will to act upon them.

TAKEAWAY

FDR COMBINED A STRONG AND RIGOROUS appeal to reason with a natural-born psychologist's understanding of the emotions in order to create powerfully persuasive appeals for unity, collaboration, sacrifice, and faith in the government. Using the pervasive medium of radio, he entered the lives of virtually every American and moved them to make the decisions he wanted them to make. The Roosevelt years may well have been democracy's finest hour, with a free people, inspired, guided, coaxed, and finally "gentled" into a series of decisions that led a broken nation toward a healing unity.

CHAPTER 5

Deciding on the New Deal

How FDR Sold the Welfare State—and Why

Previous page: *Bread line, New York, 1933.*

Deciding on the New Deal

A FORMER ASSISTANT SECRETARY OF THE NAVY in the administration of President Woodrow Wilson and an avid skipper of small sailing craft, Franklin Delano Roosevelt cheerfully confessed himself a "good sailor," but flying in an airplane was something else again. He did not take comfortably to the air. Yet when word reached him in Albany—in 1932, FDR was in his second term as governor of New York—that he had received the Democratic Party's nomination as its candidate for president, Roosevelt instantly boarded a plane and flew to Chicago.

He did not have to do it. As a matter of fact, tradition dictated that he should not have done it. Never, since parties began holding conventions to nominate candidates, had the nominee made an appearance at the convention to deliver an acceptance speech. It just wasn't done. Instead, as Roosevelt noted ironically in his tradition-shattering acceptance speech, the candidate was supposed to stage a charade of remaining "in professed ignorance of what has happened for weeks until he is formally notified . . . many weeks later." FDR decided to break tradition, both by flying—highly unusual for just about anybody in 1932 and unheard of for a political candidate—and by appearing at the convention.

He began his innovative speech with an explanation: "The appearance before a national convention of its nominee for president, to be formally notified of his selection, is unprecedented and unusual, but these are unprecedented and unusual times." He continued by explaining that, in refusing to perpetuate the time-honored nominee charade, he hoped to offer a "symbol of my intention to be honest and to avoid all hypocrisy or sham, to avoid all silly shutting of the eyes to the truth in this campaign." Moreover, he asked that his action also be taken as "symbolic that . . . from now on the task of our Party [is] to break foolish traditions. . . ."

Mired in a catastrophic economic depression, the American nation was stuck, and Franklin Delano Roosevelt understood that bringing relief to the Great Depression would require the making of many decisions to do things that had never been done before, at least not on a national scale. He therefore decided that the very first decision the American people would have to make was to discard "foolish traditions," including all outworn, unproductive, and downright counterproductive practices and values. What better way to motivate this initial decision in others than by demonstrating that he himself had made just such a decision by flying to the convention to accept his party's nomination?

FDR not only had the courage to break with tradition; he ensured that those who had chosen him as their champion were made aware that breaking with tradition was what he intended his administration to be all about. To be sure, flying to Chicago was inconvenient, unpleasant, and—in terms of precedent—uncalled for. But as a symbol, it was a decision to inaugurate a new era, to usher in what he called in the very speech he made that evening, July 2, 1932: a "New Deal for the American people."

Having so dramatically created the first node in what would be a vast network of bold decisions—each of which would ultimately require the assent and full cooperation of the American people—FDR had to sell America on the idea that not only were bold decisions required in the depths of the prevailing economic crisis, but that they were actually possible. The incumbent president, Republican Herbert Hoover, was not, as many of his critics claimed, deaf and blind to the urgent gravity of the national crisis. Indeed, he himself had proposed a series of programs and plans to deal with it. But always the line was drawn at actually providing direct federal aid to individuals and businesses. Each time he drew this line, President Hoover offered a number of arguments for withholding direct federal aid, but, as FDR saw it, the principal argument—the controlling argument—was simply that such a step had never been taken before.

"Our Republican leaders tell us economic laws—sacred, inviolable, unchangeable—cause panics which no one could prevent. But while they prate of economic laws, men and women are starving." Roosevelt continued: "We must lay hold of the fact that economic laws are not made by nature. They are made by human beings." The implication was unmistakable: made by human beings, economic laws could be changed by human beings.

Change was the core of Roosevelt's decision-making strategy. He knew that a majority of the suffering nation desperately wanted change. That powerful need was good for his program, but he also recognized that current leaders were telling the American people, in effect, that change was not possible. The first step, therefore, in implementing the strategy of change was to reverse that opinion by making everyone aware of the very real *possibility* of change. Not everything can be changed, but if a thing is made by human beings, human beings can change it.

Roosevelt came to the platform in Chicago armed with a not-so-secret weapon. That he was proposing a sharp break with tradition, as dramatically demonstrated by his very appearance at the convention, was a formidable objective. Yet he knew that he had received the nomination because, as governor of New York, he had already drawn national attention for the boldly innovative programs that had brought a significant measure of social and economic relief to his own state. The nation was already looking toward Roosevelt, who, more than any other leader in American history, had practical experience with the kind of radical change that produced tangible and positive results.

As New York's governor, FDR had shown America that he was capable of "thinking outside the box"—to use a modern cliché that was quite unknown to the Roosevelt era. It is a pity that this expression has been overused in recent years to the point that we tune it out whenever we hear it. It is, in fact, a most provocative figure of speech. Listen to it with fresh ears, and you'll find it difficult to resist the urge to crane your neck or squint your eyes to catch a glimpse outside or around that proverbial box.

But how do you go about thinking outside it?

As New York's innovative governor, Roosevelt did not really begin by *thinking* outside the box at all. Instead, he began by *rethinking* the box itself. His approach to the problems of his state began with his taking a fresh look at the elements of each issue and problem, the very elements that everyone tends to take for granted—the elements to which, ordinarily, no thought would be given at all. Governor Roosevelt reexamined them, rethought them, and reconsidered the nature, the strengths, the liabilities, the weaknesses, the purpose of just about everything a government does. Rethinking these elements allowed him to make an entirely new approach to the problems at hand. It allowed him to carry his thinking out of the box. In his annual message to the New York State Legislature on August 28, 1931, Roosevelt began by inviting legislators to ask a very basic question, a question that required them to do what he had done: first to rethink the box, and then to start thinking outside of it.

By way of introducing a radical and unprecedented program of state-funded measures to address the Great Depression, Governor Roosevelt began his speech with a startlingly basic question: "What is the State?"

And he continued by offering an answer:

> It is the duly constituted representative of an organized society of human beings, created by them for their mutual protection and well-being. "The State" or "The Government" is but the machinery through which such mutual aid and protection are achieved. The cave man fought for existence unaided or even opposed by his fellow man, but today the humblest citizen of our State stands protected by all the power and strength of his Government. Our Government is not the master but the creature of the people. The duty of the State toward the citizens is the duty of the servant to its master. The people have created it; the people, by common consent, permit its continual existence.

> One of these duties of the State is that of caring for those of its citizens who find themselves the victims of such adverse circumstances as makes them unable to obtain even the necessities for mere existence without the aid of others. That responsibility is recognized by every civilized Nation.

By reexamining and revising his thinking on what is traditionally taken for granted, Governor Roosevelt laid the groundwork in New York State for a vast program of Depression-era relief, a prelude to the New Deal he would introduce after his election to the presidency. By posing a question that prompted a reevaluation of the most basic of premises, Roosevelt began leading the nation toward unprecedented innovation.

To the delegates at the Democratic National Convention in 1932, nominee Roosevelt declared: "I pledge you, I pledge myself, to a new deal for the American people." What he went on to propose was vast in scope and included massive federal spending for direct economic relief and for an ambitious and comprehensive program of public works. Additionally, he proposed a plan to curb the agricultural overproduction that was depressing farm prices, a policy of conservation of environmental resources, a program to generate public power, a program to provide old-age pensions and unemployment insurance, a policy to regulate the stock exchange, and the repeal of prohibition. In his nomination acceptance speech, he introduced this vast and entirely unprecedented program by asking the delegates to "look a little at the recent history and the simple economics, the kind of economics that you and I and the average man and woman talk." Just as he was convinced that making the national decision for change required, first and foremost, a firm belief that change was even possible, so Franklin Roosevelt saw that the next necessary step was to ensure that everyone understood the basis of the changes that he was about to propose. He believed that people could wholeheartedly embrace only what they understood truly and deeply. To create this level of understanding, he continued his speech:

> In the years before 1929 we know that this country had completed a vast cycle of building and inflation; for ten years we expanded on the theory of repairing the wastes of the War, but actually expanding far beyond that, and also beyond our natural and normal growth. Now it is worth remembering, and the cold figures of finance prove it, that during that time there was little or no drop in the prices that the consumer had to pay, although those same figures proved that the cost of production fell very greatly; corporate profit resulting from this period was enormous; at the same time little of that profit was devoted to the reduction of prices. The consumer was forgotten. Very little of it went into increased wages; the worker was forgotten, and by no means an adequate proportion was even paid out in dividends—the stockholder was forgotten.

As FDR explained it, in the unfolding of recent economic history, the government of the United States had "forgotten" about many people. Indeed, it had forgotten about almost everyone except big business. The result? Candidate Roosevelt explained:

> Enormous corporate surpluses piled up—the most stupendous in history. Where, under the spell of delirious speculation, did those surpluses go? Let us talk economics that the figures prove and that we can understand. Why, they went chiefly in two directions: first, into new and unnecessary plants which now stand stark and idle; and second, into the call-money market of Wall Street, either directly by the corporations, or indirectly through the banks. Those are the facts. Why blink at them?

It was a speech no presidential nominee had ever made before, because, adhering to outworn tradition, no nominee had ever before appeared at a national convention to accept the nomination of his party. Breaking with the "absurd" tradition of, in effect, "blinking" at the obvious fact of nomination, FDR asked why the nation should likewise persist in blinking at the historical circumstances that had brought on the Great Depression.

You are suffering, he said. You want change. Decide first that change is possible. Then understand the reasons and the basis for change. Decide to understand the truth and face the truth.

"Then came the crash," FDR continued his speech. "You know the story."

> Surpluses invested in unnecessary plants became idle. Men lost their jobs; purchasing power dried up; banks became frightened and started calling loans. Those who had money were afraid to part with it. Credit contracted. Industry stopped. Commerce declined, and unemployment mounted. And there we are today.

For most Americans in 1932, the effects of the Depression were all too apparent. They included hunger, homelessness, fear, and despair. Yet the nature of the Depression–including, above all, its causes–was, to most, utterly mysterious and, therefore, all the more terrifying. Roosevelt sought to slice through the mystery and remove the terror, thereby putting those who heard and heeded his speech in a position from which they were empowered to make the necessary decisions for change.

Having secured the nomination, Franklin D. Roosevelt began campaigning. In a speech to the Commonwealth Club of San Francisco on September 23, 1932, he made a typically bold declaration, asserting that "every man has a right to life; and this means that he has also a right to make a comfortable living." It was a most provocative leap of reasoning, another challenge aimed squarely at questioning the very basics we take most for granted. Surely, no American would argue against the assertion in the Declaration of Independence that we all have the "unalienable rights" to "life, liberty, and the pursuit of happiness." Nor would any argue against the Bill of Rights, which guarantees that no one may be deprived of life, liberty, or property without due process of law. Candidate Roosevelt understood that these were givens—cornerstones, no less—and that the first part of his own sentence, "Every man has a right to life," would likewise be accepted as a given. In this understanding lay the core of his capacity to lead an entire nation through the matrix of decisions that would be required to build the complex and far-reaching social innovations of the New Deal. FDR had an unerring instinct for redefining and expanding the familiar in order to achieve the most radical and unprecedented reform.

He began, in this San Francisco speech, on the good, old, and familiar ground of the right to life. Then he raised upon this comfortable foundation the strikingly novel edifice of the New Deal by boldly extending the definition of that right to life to encompass the right to make a "*comfortable* living."

He took great care to avoid confusing the right to a comfortable living with a right to live off the government dole. A person, he continued his speech, may "by sloth or crime decline to exercise that right [to a comfortable living], but it may not be denied him." He pointed out that, in the Depression year of 1932, "we have no actual famine or dearth; our industrial and agricultural mechanism can produce enough and to spare." Because this was the case, "Our government, formal and informal, political and economic, owes to every one an avenue to possess himself of a portion of that plenty sufficient for his needs, through his own work."

Under FDR's incumbent opponent, Herbert Hoover, the federal government offered no direct relief from the Depression. In part, this was because Hoover and his administration truly believed the federal government had no proper constitutional authority to provide such relief, but it was also the case that Hoover and his men were simply incapable of re-imagining the federal government as a body that *could* possibly help. Herbert Hoover had come to office with a reputation as an international humanitarian, who had brilliantly administered international relief programs for a stricken Europe after World War I. Surely he—and everyone in his administration—would have agreed that all people have a right to life. Yet the Hoover administration could not have made the intellectual and imaginative leap required to entertain, let alone embrace, the redefinition of that right to encompass the right to make a *comfortable* living. Even more significantly, the administration would have been incapable even of imagining such a redefinition.

Yet, Roosevelt was confident, once the decision was made to redefine that right, that all the other decisions required in the federal government's matrix of decision would suddenly become not just thinkable, but feasible—and not just feasible, but obligatory. Decide to redefine the right to life as a right to make a comfortable living, and the federal government suddenly acquires a vast new set of obligations to meet, each requiring decisions on how to meet them.

Franklin Delano Roosevelt defeated Herbert Hoover in 1932 and was inaugurated on March 4, 1933, at the very nadir of the national economic crisis, when one in four Americans was unemployed. The failure of banks and other financial institutions had created widespread panic, which, in turn, brought about the collapse of even more banks. Roosevelt entered this dismal and dangerous picture by projecting an image of realistic and believable confidence and optimism. His assertion in his inaugural address, that "the only thing we have to fear is fear itself—nameless, unreasoning, unjustified terror," rung true precisely because he promised leadership, pledging to do far more than merely stand by as the Depression deepened. He would, he told the American people, obtain from Congress "the one remaining instrument to meet the crisis—broad executive power to wage a war against the emergency, as great as the power that would be given to me if we were in fact invaded by a foreign foe."

Almost immediately on taking office, Roosevelt declared a "bank holiday" to stave off a panic run on those financial institutions that were still functioning. Then he called Congress into emergency session. During the first "Hundred Days" of his administration, FDR introduced a sweeping program of relief measures. He took the nation off the gold standard, a step that provided a measure of relief to debtors and exporters. He prevailed on Congress to appropriate $500 million in federal relief grants to states and local agencies. Under Harry Hopkins, the former social worker who had been his right hand during his terms as New York governor, FDR created the Federal Emergency Relief Administration, the Civil Works Administration (CWA), the Civilian Conservation Corps (CCC), the Home Owners Loan Corporation (HOLC), and the Public Works Administration (PWA). The CCC employed more than 2.5 million young men in conservation work, the HOLC furnished emergency assistance to mortgagors and homeowners, enabling them to avoid foreclosure, and the PWA created great public works projects.

These and other programs and initiatives not only provided immediate relief, but also restored hope among those hardest hit by the Depression. Without such measures, many believed the nation would have plunged into civil insurrection, or even outright revolution.

The New Deal was structured on a vast and all-encompassing scale. In addition to emergency relief, it included a program of long-range reform. The Federal Deposit Insurance Corporation (FDIC) insured bank deposits, an important safeguard against runs. The Securities and Exchange Commission (SEC) introduced the regulation of the stock exchanges, a safeguard against the kinds of practices that had contributed to the sudden fall of the markets in 1929. The Tennessee Valley Authority (TVA) built great multipurpose dams to control floods and generate cheap hydroelectric power. Two agencies of special importance were the National Recovery Administration (NRA) and the Agricultural Adjustment Administration (AAA). The NRA provided incentives to management as well as labor to establish codes of fair competition within each industry, including equitable pricing and production policies, collective bargaining, minimum wage rules, and regulations on setting maximum working hours. The AAA sought to raise farm prices by setting production quotas approved by farmers in referenda and subsidizing farmers who stayed within the quotas.

Historians and economists continue to debate the overall efficacy of the New Deal. No one believes that it alone ended the Great Depression, but most historians agree that, whatever its successes and failures, it did provide quick relief, emergency aid, and most important, long-term hope.

The leadership task Roosevelt faced had been supremely difficult, not just from an emotional and economic perspective, but ideologically as well. He had to sell decisions that took the nation on a perilous course between preserving democracy, capitalism, and a free market economy on the one hand, and providing intervention, support, and outright control on the other. In 1935, FDR ushered through Congress three of his most sweeping initiatives: the Works Progress Administration (WPA), which employed millions in work relief programs; the Wagner Act, which set up the National Labor Relations Board (NLRB) and thereby guaranteed labor the right to bargain collectively on equal terms with management; and Social Security, which provided for federal payment of old-age pensions and for federal-state cooperation in support of unemployment compensation and relief for the blind, the disabled, and families with dependent children.

Leadership is as much about continuity as it is about change. Facing an economic crisis unprecedented in duration and extent, Roosevelt accepted the need for radical changes in government, yet he also understood that introducing too great a degree of change during a crisis would only deepen the people's sense of crisis. His leadership objective therefore defined the necessary change within a familiar framework. In his Second Annual Message to Congress on January 4, 1935, he explained: "We have undertaken a new order of things; yet we progress to it under the framework and in the spirit and intent of the American Constitution. . . . We seek [change] through tested liberal traditions, through processes which retain all of the deep essentials of that republican form of representative government first given to a troubled world by the United States."

To lead simultaneously for change and for continuity requires sensitivity and finesse. There is a great risk of conveying an ambiguous message or of suggesting that one is ambivalent—and nothing is more fatal to effective leadership than ambivalence. Even worse, there is the danger that the message will come across as so much double-talk, as if what is being offered is, in effect, something entirely new that's worked well in the past. Roosevelt avoided these traps by making absolutely clear, in every speech he delivered, that the change he proposed not only lay within the parameters of the Constitution, but actually grew directly out of it, the very document on which the nation was founded. Roosevelt broke with outworn tradition by appealing precisely to the most valued, venerable, and useful tradition. In this way, he made it easier for the American people to decide to accept the New Deal.

THAT THE AMERICAN ELECTORATE GENERALLY APPROVED of the program was evident in Roosevelt's landslide reelection in 1936. In typical FDR fashion, however, he did not accept this victory as a cause for complacent self-congratulation. He faced the American people squarely in his second inaugural address and issued a bold challenge to himself and to the New Deal. He declared: "I see one-third of a nation ill-housed, ill-clad, ill-nourished." In this statement, he acknowledged the scope and difficulty of the ongoing problems as well as the absolute need to press on with continued reform, relief, and recovery. Roosevelt never stopped selling the welfare state because he understood that its continuance and its prospects of ultimate success depended on the American people deciding to support it as each and every day dawned. A dynamic system, the New Deal could never be an accomplished fact. It was perpetually a work in progress, fueled by the ongoing decision of the people to support it.

TAKEAWAY

ROOSEVELT'S GREAT STRENGTH as a leader was to unite clarity with confidence. In 1933, when he took office during the worst of the Great Depression, the American people were gripped by confusion and fear. FDR brought clarity by defining the situation in terms of the "Five Ps" we enumerated in Chapter 1: he presented the Problem "unblinkingly," he defined where he and the American people stood in relation to it—their Position, he presented the Possibilities this problem and position offered, and, in the New Deal, he delivered a Proposal to deal with the problem. Always he added a full and frank acknowledgment of the final P, the Price, the sacrifices that would be required to resolve the problems presented by the current situation. Having brought order to confusion, FDR next vanquished fear by instilling in everyone who heard him the confidence that, by working together in good faith, the possibilities could be realized, the proposal would succeed, and the problem would be solved.

CHAPTER 6

Island Hopping

The Pacific Strategy That Won World War II

Previous page: *U.S. troops go over the side of a Coast Guard–manned combat transport to enter the landing barges at Empress Augusta Bay, Bougainville, November 1943.*

Island Hopping

THE EXPLOSION OF WORLD WAR II in the Pacific was catastrophic for the United States and its allies. In his war message to Congress on December 8, 1941, the day after the Japanese attack on Pearl Harbor, President Franklin D. Roosevelt delivered an unblinking summary of the situation. After reporting that the Pearl Harbor attack had caused "severe damage to American naval and military forces" and "that very many American lives have been lost," the president launched into a terrible catalogue of enemy conquest:

> In addition American ships have been reported torpedoed on the high seas between San Francisco and Honolulu.
>
> Yesterday the Japanese government also launched an attack against Malaya.
>
> Last night Japanese forces attacked Hong Kong.
>
> Last night Japanese forces attacked Guam.
>
> Last night Japanese forces attacked the Philippine Islands.
>
> Last night Japanese forces attacked Wake Island.
>
> And this morning the Japanese attacked Midway Island.

In the days, weeks, and even months that followed FDR's war message, Japanese forces overran and captured even more Asian and Pacific territory in their far-reaching campaign to control the food and raw material resources of war and to build around their empire a vast defensive perimeter intended to insulate and defend Japan's home islands, ensuring that the war they had begun would never come back to touch them. In February 1942, as part of this Pacific campaign, Japanese forces captured Rabaul, on New Britain Island, Territory of New Guinea, which had been held by an outnumbered garrison of the Australian army. Rabaul was instantly transformed into Japan's most formidable forward army, navy, air, and supply base in the South Pacific. It was a fortress, a bastion that dominated both the South and Central Pacific theaters. Before the Allies could even begin to effectively counter the Japanese juggernaut, let alone attempt to turn the tide against Japan, Rabaul would have to be dealt with, which meant attacking, overrunning, and capturing this most thoroughly defended of objectives.

That much seemed self-evident. After all, what else could be done with the enemy's most powerful and strategically placed position on the battlefield?

Certainly, General Douglas MacArthur, the supreme Allied commander in the theater, thought of the situation this way. He drew up the "Elkton Plan," a strategy for launching an attack on Rabaul from Allied bases in Australia and New Guinea with the object of overrunning the base and taking it. His U.S. Navy counterpart, Chief of Naval Operations Admiral Ernest J. King, drew up a similar plan, which, however, put the navy rather than the army in the forefront of operations. Even in the midst of a desperate war, the army and navy were keenly–and not always productively–competitive, but what neither MacArthur nor King took into account was the larger matrix in which the Allied leaders were making their war decisions. Both President Roosevelt and British Prime Minister Winston Churchill were determined to focus the major Allied effort against Adolf Hitler and Benito Mussolini in Europe, leaving the war against Japan in a more or less defensive holding pattern until the European situation could be contained. U.S. Army Chief of Staff General George C. Marshall, acting in accordance with this larger strategy, offered a compromise plan for the Pacific, in which both the navy and the army would play leading roles. First, the navy would lead campaigns to capture Tulagi, Guadalcanal, and the Santa Cruz Islands in an operation dubbed Watchtower. Next, the army would take the lead in attacking and capturing the northeast coast of New Guinea and the central Solomon Islands. As for Rabaul, the mission as Marshall's plan reconceived it was no longer to capture the enemy bases there, but merely to reduce them, neutralizing their effectiveness and knocking them out of the war.

The revised plan resulted, first and foremost, in a very long battle for Allied possession of Guadalcanal, which was followed by the rapid occupation of the Russell Islands by the end of February 1943. Now finding themselves on the defensive, the Japanese sent reinforcements to the area, which resulted in the Battle of the Bismarck Sea (March 2–4, 1943), in which MacArthur's air arm inflicted severe losses on the Japanese. The Japanese theater commander, Admiral Isoroku Yamamoto, responded with

"I-GO," a counteroffensive operation aimed at retaking Guadalcanal and New Guinea. MacArthur succeeded in beating this off, and hoped that he would now be permitted to launch a new version of the Elkton plan, "Elkton III," again aimed at capturing Rabaul.

Elkton III called for MacArthur to attack northeast New Guinea and western New Britain Island while Vice Admiral William F. "Bull" Halsey, commanding the South Pacific Area, hit the central Solomon Islands. It was a tremendously ambitious undertaking, which, on the ground, required seven divisions to be added to those already in the theater. The British protested what they deemed an extravagant use of manpower, prompting the U.S. Joint Chiefs of Staff to insist on a compromise. Elkton III would be delayed two months, and then would use forces already in theater or en route to it. This modified plan was called Operation Cartwheel by the military, but the civilian public of the Allied nations came to call it the "island-hopping strategy." By either name, it guided virtually all of the strategic decisions that the Allies made in the rest of the Pacific war.

Operation Cartwheel was a two-pronged advance through the South West Pacific Area (SWPA), under the command of MacArthur, whose forces moved along the northeast coast of New Guinea and occupied the islands nearby, and Admiral Chester W. Nimitz, whose forces advanced through the Solomon Islands toward Bougainville. As originally conceived, Operation Cartwheel included a total of thirteen sub-operations, aimed at taking various islands, all with the main object of neutralizing Rabaul not by attacking it immediately or directly, but by isolating it from supply and reinforcement. Once they were under way, the campaigns and battles of Operation Cartwheel proved at once more effective as well as more costly than anticipated.

Island by island, the Japanese were defeated, but they did not give up. In each instance, over and over, victory did not come with the surrender of Japanese soldiers, it came only with their deaths. No battle was won until virtually all of the enemy lay dead. For this reason, in August 1943, when President Roosevelt met with Prime Minister Churchill in Quebec for the Quadrant Conference, which took place while Cartwheel was well under way, the decision was made to bypass Rabaul entirely, as well as two other objectives, Kavieng, capital of New Ireland province in Papua New Guinea, and Kolombangara, a Japanese-held island in the New Georgia Islands group of the Solomons. The decision to leap over these objectives was by no means a retreat or a lowering of expectations, but was, in fact, the result of a stunning strategic insight. Allied planners realized that the Japanese had been able to occupy so many Pacific island outposts by transforming them into a vast network for communications and supply. That was a great strength, of course, but, Allied planners now saw, it was also a fatal weakness. Capture just the right islands in the network, and the others would be cut off and thereby rendered useless. For military purposes, they would be left to wither on the vine. Attacking head on so formidable a concentration of troops and equipment as were at Rabaul would be terribly costly. Why not, therefore, just cut them off? By hopping over this objective and proceeding to the next, Rabaul would not be destroyed or captured, but it would be utterly neutralized, which was just as effective and a lot cheaper.

At first, the Quadrant planners instructed MacArthur to target the Japanese base at Kavieng instead of Rabaul, but, on further thought, soon realized that this objective, too, could be hopped over and left to wither, along with Kolombangara. Thus the roster of Operation Cartwheel sub-operations was reduced from thirteen to ten: in Operation Chronicle, to be launched on June 30, 1943, U.S. Army forces would assault Woodlark Island and Kiriwina Island (in the Trobriand Islands group). On the same day, Operation Toenails would step off, using U.S. Marine and U.S. Army forces and aimed at taking New Georgia Island and Rendova. Operation Toenails would continue on July 5, with attacks on the towns of Zanana and Bairoko, New Georgia Island, then against Vella Lavella on August 15 and Arundel Island on August 27. The next Cartwheel sub-operation, Operation Postern, would target Lae, New Guinea, on September 5 and involve two Australian divisions supported by the U.S. 503rd Parachute Infantry Regiment. On October 27, Operation Goodtime would be launched against the Treasury Islands, using the New Zealand 8th Brigade. The U.S. 2nd Marine Parachute Battalion would be responsible for carrying out Operation Blissful, to capture Choiseul Island, beginning on October 28, and, on November 1, a division of U.S. Marines and U.S. Army forces would launch Operation Cherryblossom against the formidable Papuan island of Bougainville. This would be followed on December 15 by Operation Dexterity, which involved assaults on Saidor, New Guinea, and an amphibious landing at Cape Gloucester, New

Britain Island, as well as an assault on Arawe, New Britain. On the day after Christmas, the 1st Marine Division would reinforce the assault on Cape Gloucester, and on January 2, 1944, the U.S. 32nd Infantry Division would be added to the forces attacking Saidor. The Admiralty Islands would be assaulted by U.S. 1st Cavalry Division beginning on February 29, followed by a marine assault on Emirau Island on March 20.

Even with three major objectives designated to be hopped over, Operation Cartwheel was an ambitious campaign, and Douglas MacArthur, a stubborn commander who did not readily bow to the strategy of others, at first resisted abridging it in any way. On further reflection, however, he embraced island hopping, having realized that, by resetting the objective from occupying to neutralizing Rabaul, Kavieng, and Kolombangara, he was achieving the results intended for his pet plan, Elkton III. Accepting this result eagerly, he directed the invasion of Saidor, then, while the other Cartwheel sub-operations were under way, he commenced Operation Reno, an advance across the northern coast of New Guinea to Mindanao, in the southern Philippines—a first step toward redeeming the pledge he had made at the beginning of the war, after following presidential orders to turn over his Philippine command to Lieutenant General Jonathan Wainwright and, with his family and immediate staff, evacuate for Australia. After that hazardous flight, he broadcast to the Philippines and the world: "I shall return."

Island hopping was a brilliant departure from the kind of linear thinking that all too often characterized the strategic policy of the U.S. military, which tended to emphasize head-on attacks relying primarily on main strength rather than agile, indirect tactics. With island hopping, the Americans enlisted and exploited as a key ally against the Japanese the logistical pressures created by vast Pacific distances. Cut the principal links in the chain of Japanese conquest, and supply and communication were virtually impossible. Allied air and submarine attacks were used extensively to sever the lines of supply and communication between the islands, and only selected positions were directly contested and occupied. In this way, progress toward the Japanese home islands was greatly accelerated. In the northern sector of Operation Cartwheel, American forces advanced from Midway into the central Pacific, reaching Iwo Jima—a sentinel position guarding the Japanese home islands—in February 1945. In the southern sector, Allied forces used hard-won Guadalcanal as a springboard from which to advance into the Solomon Islands and, from here, to a massive invasion of the Philippines by early 1945.

CONQUEST OF A CONTINUOUS CHAIN OF ISLAND positions obligated the Japanese to maintain a hold on each link of the chain. Possession of so much Pacific real estate was intimidating to the Allies, until the realization dawned on Allied planners that defending these vast holdings was a daunting task that rendered the mighty Japanese conquerors highly vulnerable. The Allies' decision to use the fruits of Japanese conquest against the conquerors themselves was a brilliant stroke of strategy on a theater-wide scale. It saved a great deal of time and untold thousands of American and other Allied lives in battles that did not have to be fought and never were.

TAKEAWAY

"ISLAND HOPPING" WAS THE STRATEGIC equivalent of jujitsu. It exploited the supposed strength of the opponent's position—the vast Japanese network of conquests—to defeat that opponent. It was RISK played out in the real world: a magnificent visualization of winning, big-picture strategy, which turned all of the enemy's assets into liabilities. Overwhelmingly, war is characterized by waste, extravagant and vast. In the island-hopping strategy, the American strategists had found a remarkably economical means of turning the tide in the Pacific struggle.

CHAPTER 7

The FedEx Idea

Why Overnight Shipping Works

Previous page: *An engraving of carrier pigeons, with messages attached, from* Harper's New Monthly Magazine, *April 1873.*

The FedEx Idea

THE ROMANCE OF AMERICAN HISTORY is largely the romance of space, of the vast, beautiful, difficult, dangerous voids that separated, through most of the continent's history, one pocket of settlement from another. And the romance of American space has long been the story of moving through that space, engaging it, and conquering it—from the sixteenth-century wanderings of such conquistadors as Coronado, De Soto, and Cabeza de Vaca, to the Americans Lewis and Clark in the early nineteenth century, to such spiritually driven enterprises as the epic Mormon Trek some years later. There were also a number of less epic but no less intrepid commercial endeavors, including the mountain men of the fur trade, the gold rush forty-niners, the riders of the Pony Express, and the immigrant laborers who built the transcontinental railroad. In the twentieth century, there were the pioneers of air mail between the world wars, and even the homely vocation of overland trucking took on something of a gallant outlaw aura well into the 1970s.

There can be no denying that shipping or getting a piece of merchandise, a crucial spare part, or a vital document from point A to point B is the lifeblood of business and essential to commerce, to the economy, and to civilization itself. Shipping exists to transport our treasure, and in turn we pour a portion of our treasure into the great enterprise of shipping so that we can transport it farther, faster, and more reliably. The posting of money against time and space is a human undertaking worthy of the pen of an epic poet, a great novelist, or a distinguished historian. Thus, we have many accounts of the conquistadors and their ilk, of Lewis and Clark and those who followed them, of the Mormon trekkers and other spiritual pilgrims, and of the Union Pacific and the railroads that came afterward. All of these poems, novels, and histories document and celebrate the act of carrying something from somewhere to somewhere else. Whatever the form, the story is basically the same: what happened in the space between the somewheres.

In 1971, a man named Frederick W. Smith gambled his sizable inheritance on the idea that the new story—the *real* story—lay not in the physical journey from one somewhere to another, but in the intellectual strategy that determined the direction of each of those journeys. Up to this point, practically all the stories, all the energy, and all the money had been devoted to the middle phase of the journey. No one seemed to pay much attention to how a journey started. There was apparently little romance in the means of making sure Package A intended for Destination Z was not put on the truck bound for Destination B. The movement of wheels on roads or rails, of ships along rivers or across oceans, and of airplanes through clouds—these were all thrilling. Sorting the packages that rolled or sailed or flew was, on the other hand, merely a matter of routine and not worth thinking about.

But Smith decided to think about it.

He came from a transportation family: grandson of a steamboat skipper, son of a bus executive. When he was in his teens Smith learned to fly, and in the 1960s he helped put himself through Yale University, where he studied economics, by flying weekend charters. A regular customer of the company he flew for was IBM, which couldn't wait for scheduled passenger airlines to get important spare parts to customers. Smith took note, and in 1965 he wrote a term paper for one of his economics classes in which he argued that a dedicated air express delivery service would not only make money, but that it was exactly what emerging high-tech segments of corporate America most needed—and did not have.

There is a story that Smith's professor gave his paper a C, but when asked about this by a *BusinessWeek* writer in 2004, Smith said he remembered doing better than that. He ascribed the tale to urban legend. It really didn't matter, however. He could have gotten an F, and he would still have believed in the future of overnight air delivery. "I knew the idea was profound," he told the writer.

Smith served two Vietnam tours as a marine, then returned to the States and civilian life in the late 1960s. He took up his air-express delivery idea, raised $80 million from venture capitalists, put up $4 million of his own—all the money his late father had left him—and started Federal Express Corporation in 1971 at Little Rock, Arkansas. It was a very ambitious name, deliberately chosen to evoke the nation-spanning scope of the enterprise and to persuade the *actual* federal government to award some lucrative contracts. But an impressive name was not sufficient to get Smith's company over its first hurdle, the refusal of Little Rock airport administrators to provide facilities for Federal Express aircraft. Undaunted, the young entrepreneur left Little Rock for Memphis, Tennessee, and on April 17, 1973, the company made its first flights. Total cargo that night included just 186 packages.

Measured by volume, it was hardly a promising start. But it was better than what had happened little more than a month earlier on March 12, 1973, which Smith had intended as the official launch of service. That night he had a grand total of seven packages—and, to carry them, fourteen brand-new Dassault Falcon 20s, making Federal Express the first cargo airline with an all-jet fleet. By the time he relaunched in April, Smith had a network of twenty-five U.S. cities, more than twice the eleven cities he had in March. He was now flying a lot more than 186 packages a night, but volume was still way too low to sustain the business.

His pledge of guaranteed overnight delivery between any two points covered by the company's network should have kindled instant excitement and skyrocketing demand. Dedicated air express was innovative, and so was an all-jet cargo fleet. But the hub-and-spoke network idea was nothing short of revolutionary, reinventing the entire shipping process.

The choices conventionally available to shippers included freight companies, small-package delivery services, the U.S. Postal Service, special couriers, proprietary delivery systems, and miscellaneous ad hoc solutions.

Freight companies provided regularly scheduled, relatively inexpensive service, but it was typically slow and limited by the territory served by the particular freight company. If a shipper's package was being delivered outside of one freighter's territory, it had to be transferred to another carrier and perhaps another mode of transport—from truck to train, say, then back to truck again. Moreover, for the sake of economy, a shipper's package was batched with packages shipped from many others. This added time for sorting and handling and, often, for making multiple transfers from one vehicle to another. The system worked, but it consumed time.

Small-package delivery services, such as the venerable United Parcel Service (UPS), were more adept at handling special shipments, but because they still put the focus on transportation rather than sorting, the processes for ensuring that a given package arrived at its destination were usually time consuming. Moreover, these carriers relied on trucks—not aircraft—for routine shipments, so transit times were also relatively slow and, perhaps worse, variable. They were usually more predictable than shipment times for larger freight deliveries, but they still varied by distance and traffic, often moving more slowly during times of peak demand.

The U.S. Postal Service provided very cheap service and, because of its immense infrastructure, it could carry letters and small parcels anywhere. Moreover, postal service was very convenient for the shipper. Depending on what he was sending, pickup and delivery were door-to-door, mailbox to door, or, in the worst case scenario, post office to door. As with other carriers, however, delivery times tended to be variable. Even worse, for the most economical classes of mail, there was no efficient way to track a shipment. You put your letter, document, or package into the system and hoped it would get to where it was going. In fact, it usually did, but by the late 1960s and early 1970s, there was a growing perception that the postal service was becoming inefficient and careless, that a disproportionate number of shipments were delayed or lost or stolen. The perception of a system in crisis was certainly exaggerated, but it was at least partially justified. Both shippers and consignees felt that the postal service delivered second-rate service at best. If you did business with a company that used the U.S. Postal Service to ship merchandise and documents, you had a sneaking suspicion that the company was taking your patronage for granted. Between variable delivery times, the lack of a tracking system, and the feeling of being treated as unimportant, shipping via USPS came to feel like a crap shoot.

If you had something very important to send—or you wanted to impress a key customer—you could call on a courier service, which would effectively take custody of your documents or packages and carry them from your door to that of your customer, often chartering aircraft to do so, as Fred Smith had discovered as a young pilot. Couriers provided relatively high levels of security, often including fully bonded service, and tracking presented little or no difficulty. They were, however, expensive, and limited in the volume and types of shipments they could handle. Moreover, territories were often also restricted, and as with most other carriers, the cost depended in large measure on distance.

Companies that did a high volume of shipping often developed proprietary delivery systems. Catalogue companies and big merchandisers maintained their own warehouses and truck fleets. These were excellent solutions, if you delivered enough product to warrant the high capital costs. For small businesses—and even for most medium-sized businesses—proprietary solutions were not an answer. Indeed, "fulfillment"—warehousing, distribution, and delivery—was, for many prospective start-up companies, a more formidable hurdle than research and development, or even manufacturing.

Searching for a fulfillment strategy that would allow them to enter the marketplace, entrepreneurs began to turn to a business model dubbed "Just In Time"—JIT—a streamlined approach to inventory. Traditionally, manufacturers maintained a stockpile inventory of needed parts, and retailers maintained similar inventories of merchandise. Doing this required a capital investment in the parts or merchandise as well as in warehouse facilities. Neither of these, of course, represented a rapid return on investment. A large inventory was dead money. The origin of JIT may be found in the origin of mass production itself. In his *My Life and Work*, published in 1922, Henry Ford explained that we "found in buying materials that it is not worthwhile to buy for other than immediate needs." That sounds simple enough—just good common sense. But Ford continued: "We buy only enough to fit into the plan of production, taking into consideration the state of transportation at the time. If transportation were perfect and an even flow of materials could be assured, it would not be necessary to carry any stock whatsoever. The carloads of raw materials would arrive on schedule and in the planned order and amounts, and go from the railway cars into production. That would save a great deal of money, for it would give a very rapid turnover and thus decrease the amount of money tied up in materials." But, of course, transportation was not perfect early in the twentieth century any more than it was in 1973, when Federal Express made its first flights. As Ford noted, "With bad transportation one has to carry larger stocks," and so, for Henry Ford, true JIT remained an unattainable ideal.

Provide really, really good, reliable, flexible transportation, and JIT becomes a viable option. With JIT a reality, small and medium-sized businesses suddenly are empowered to do what only big business could do before.

The Federal Express model collected shipments from individual shippers, brought them to a company-owned sort facility, from which they were transported to an airport-based hub (either the main Memphis hub or a smaller hub in one of the network cities), which routed the shipment onto the appropriate aircraft. On landing, the shipment was received and routed in another airport-based hub and sent either to a local Federal Express facility or trucked directly to the consignee. Effort was focused on developing the sorting and routing processes to optimum efficiency. In effect, Federal Express, ostensibly a transport company, served as the shipper's warehouse even as super-streamlined sorting and routing procedures ensured that the hubs would never actually warehouse the shipments, but would push them through—with dead-on accuracy—as fast as possible.

Since identifying and tracking shipments was key to sorting and routing them efficiently, tracking also became an added customer benefit. As Federal Express grew its business, this benefit became a real drawing card. The product *feature* Federal Express sold was rapid and accurate door-to-door delivery. The product *benefit* became peace of mind. No longer did a shipper have to feel that consigning his precious package to a transport company was an act of faith. With the ability to track the shipment, to keep tabs on it almost in real time, he felt that he had not relinquished control over it. To what became the company's widely recognized advertising slogan, "When it absolutely, positively has to get there overnight," another line, "Relax, it's FedEx," was added to tout the product's chief benefit.

In 1975, the company introduced convenience to the sense of confidence it offered. In this year, the first "Drop Box" was installed. Suddenly, sending something by Federal Express became as easy and routine as dropping a letter in the traditional mailbox, and this began to close the gap between the premium-priced private shipper and the less expensive U.S. Postal Service.

Smith and his backers saw all of the company's potential, but in its start-up, Federal Express struggled against a host of daunting obstacles. Volume, initially very weak, was slow to grow. As if that weren't enough to kill the fledgling company, Federal Express found itself squeezed between the U.S. postal monopoly on the one hand, which prohibited Federal Express from delivering documents until the regulations were changed in the late 1970s, and on the other hand, Federal Aviation Authority (FAA) and airline regulations, which initially restricted the company to operating small jets that could handle only small-volume shipments and certainly no major freight. The Dassault Falcon 20s were essentially small business jets. Configured for passengers, each could carry just eight to ten. Set up for cargo, the aircraft could carry 2.5 tons, no more than the capacity of a small truck.

Fortunately for Federal Express, the transition of the American economy from heavy manufacturing to service and high-tech information industries was producing a lot of small parcels that needed to be delivered quickly and reliably. The explosion of the computer and electronics industries, in particular, made a resounding financial impact on the new company. Nevertheless, at one point in 1973, Federal Express was so strapped for cash that Smith flew one of his planes to Las Vegas, where he hit the blackjack tables, won $27,000, and wired it promptly to Memphis.

The ability of the company and its founder to simply hang on had paid off by the end of the 1970s. New postal regulations at last allowed Federal Express to deliver documents, and in 1977, after two years of relentless congressional lobbying, Public Law 95-163 was enacted, which enabled Federal Express and other cargo airlines to operate larger aircraft without geographic restrictions on routes. The company quickly bought seven Boeing 727s, each of which had a cargo capacity of twenty tons—eight times that of the Dassault Falcon 20. By this time, the development of computer technology was making sorting and routing ever more efficient. In 1979, the company launched its Customer Operations Service Master Online System (COSMOS) centralized computer system, which enabled the real-time coordination and management of personnel, packages, vehicles, and weather scenarios. This was supplemented the following year by Digitally Assisted Dispatch System (DADS), a technology that allowed for the efficient real-time coordination of on-call pickups for customers.

In 1981, Federal Express went international by offering service to Canada, and in 1984, after acquiring Gelco Express International, it launched Asian and Pacific operations. During this period of early growth, company management also recognized a parallel revolution taking place in computers and information technology. Just as the Federal Express model had enabled small and medium-sized businesses to liberate themselves from the need of capital-intensive central warehouses, so the emergence of the personal computer (PC) was liberating businesses and individuals from massive mainframe computer systems and their high capital costs. In 1984, Federal Express introduced a PC-based automated shipping system, which enabled users to prepare shipments, order service, and request pick-up from their desktop computers. Around the same time, the company introduced machine-readable bar code labeling and, in 1986, the SuperTracker, a hand-held bar code scanner. Thus, a package could be instantly tracked at every stage of the shipping operation, from pick-up to delivery. In 1989, Federal Express extended service to Europe and established a satellite tracking system by which company delivery vehicles could be pinpointed. Computer, communications, and tracking systems were further developed during the 1990s and into the twenty-first century.

In 1994, the company even streamlined its corporate name, condensing it from Federal Express to FedEx. It also launched fedex.com, the world's first transportation Web site, which gave shippers the capability of doing all of their FedEx business online, including tracking and payment. As the Web rapidly changed the way the world did business, spawning the cyber realm of e-commerce, FedEx partnered with major online merchants to handle many aspects of order fulfillment more directly. Back in the 1970s, the strategic focus on a sorting hub radically streamlined the shipping process by reducing the number of physical hand-offs involved in getting something from one point to another. Now, by integrating itself into online commerce, FedEx made it unnecessary for a shipper even to prepare his merchandise and call for a pick up. Instead, the order came directly to the transport company, which dispatched the appropriate vehicles for pick-up. By the early twenty-first century, the vast matrix of manufacturing, retailing, and shipping had come astoundingly close to the entirely theoretical level of perfection Henry Ford had imagined in 1922.

FedEx has many competitors, but it remains the largest air-express operator, with a 44 percent share of the market. It operates nearly 700 aircraft, carrying an average of 5.5 million shipments daily. Most of its domestic overnight freight is sorted and routed through its Memphis "Superhub," but it also operates other U.S. hubs in other cities, and an Asian-Pacific Superhub in Guangzhou Baiyun International Airport, Southern China, opened in 2008.

ALONG WITH THE BEST OF ITS COMPETITORS, FedEx has reinvented the strategy of shipping and, in doing so, reinvented commerce itself by putting logistics (the flow of needed materials within a company or between related businesses) and fulfillment (the shipping of merchandise to external customers) at the very heart of the entire commercial enterprise. Shipping is no longer just a matter of carrying something between somewhere and somewhere else. It is about information exchange and processing. It is about multiplying efficiency and responsiveness by replacing capital-intensive warehousing operations with JIT technologies and techniques. The creation and operation of systems that sort and route millions of packages and documents daily, yet still deliver them one at a time and without error, is as much about brainpower as horsepower.

TAKEAWAY

THE FEDEX STORY BEGAN WHEN Frederick W. Smith re-imagined the third of the decision matrix's "Five Ps," Possibilities. He saw that transportation could become more than a link between production and sales. It could be an integral phase of a seamless continuum of commerce. Moreover, he re-imagined the Possibilities of transportation itself by focusing on sorting and processing shipments rather than simply moving them. The changes made in the middle term of the Five-P process ramified throughout the decision matrix and created new ways of doing business. Indeed, the changes enabled the invention of entirely new businesses.

CHAPTER 8

Triangular Diplomacy

Nixon, Kissinger, and the World

Previous page: *Richard Nixon with Chou En-Lai in Beijing, February 25, 1972.*

Triangular Diplomacy

The United States emerged from World War II as the most powerful nation on earth, and Americans—at all levels, from the man and woman on the street to the leaders of government—rapidly became accustomed to this status, even as the nation rushed to reduce its armed forces. As the postwar euphoria was replaced by the mindset of the Cold War, in which the United States stood in competition with the Soviet Union for the role of dominant world power, Americans became accustomed to thinking of their nation as still powerful, but in a world that had become essentially bipolar. It was us versus the Soviets. This contest was played out diplomatically as well as militarily, both through a nerve-wracking, economy-draining, and potentially mass-suicidal nuclear arms race and through a series of "brushfire"—as well as larger—wars in the Third World. Some were fought between "proxy" nations or factions, one side backed by the Soviets, the other by the Americans. Other conflicts—most notably the Korean War and the Vietnam War—directly involved U.S. forces, while the Soviets managed to stay out of direct participation with ground forces, confining military support to supplying finance, arms, missiles, and some pilots who flew MiG missions.

During his rise to political prominence, first as a U.S. representative, then as a senator, and finally as vice president under Dwight D. Eisenhower, from the late 1940s to 1961 Richard M. Nixon was a classic "cold warrior," who believed that *the* enemy of the United States was the Soviet Union, and that the struggle between that superpower and

America was essentially ideological and idealistic. The United States was the avatar of democratic capitalism, whereas the Soviet Union was the embodiment of totalitarian communism. The world of Richard Nixon was the bipolar world of the early phase of the Cold War.

By the time Nixon made his second run for the White House in 1968—he had been defeated in his first bid for the presidency by John F. Kennedy in 1960—the world had changed. The Soviet Union had effectively achieved parity with the United States as a nuclear power, but it was also the case that the Soviet-dominated "Communist bloc," which had come into being shortly after World War II, was no longer the great monolith it had been at its height in the early to mid-1950s. Indeed, by 1968, relations between the Soviet Union and the People's Republic of China, the two biggest communist states, had deteriorated so badly that the nations were routinely shooting at one another across their borders over territorial disputes. In addition, former Soviet satellite nations, as well as more distant Soviet client states in the Third World, were becoming increasingly independent ("nonaligned") and, in some cases, increasingly powerful. The world was no longer divided neatly between the two superpowers. Political, economic, and military power was becoming multipolar. In contrast to many of his fellow cold warriors, presidential candidate Nixon had changed his world view to reflect the new reality. He had come to realize, as he would later put it, that the United States was moving from a position of "predominance to one of partnership." This was, in some ways, unnerving for a country accustomed to being on top, but Nixon also realized that the next president would have a historic opportunity to restructure the international order for greater stability and prosperity on both a national and international level.

In 1968, Nixon defeated Democratic contender Hubert Humphrey, in large part on the strength of an obscure pledge that he had a plan to end the Vietnam War by bringing "peace with honor." In fact, the Vietnam quagmire would figure importantly in the diplomacy Nixon formulated and implemented with Henry Kissinger, the former Harvard professor of political science who served the president first as national security advisor and then as secretary of state. Together, Nixon and Kissinger developed the concept of "triangular diplomacy," which, in conjunction with two additional diplomatic principles—what they called "realism" and "linkage"—formed the basis of the innovative foreign policy of the Nixon years. Both guided and driven by triangular diplomacy, realism, and linkage, Nixon and Kissinger strove to master and manipulate the decision matrix of world politics, making decisions themselves but—just as important—maneuvering other nations, especially the Soviet Union and China, into making decisions that would ultimately enhance the position of the United States. It was the kind of diplomacy the British in the nineteenth century had dubbed the "great game," complex and manipulative on a scale and with a deliberate strategic self-consciousness that had not been seen since the days of Germany's legendary "Iron Chancellor," Otto von Bismarck.

Because the evolution of the Nixon-Kissinger strategic decision making is complex, it helps to begin with a snapshot overview of what the pair was aiming at. Both men began by facing the reality that the confident and comfortable days of assured American military superiority were at an end. Yet also at an end were the days in which the communist powers—chiefly the Soviet Union and China—were the inevitable and implacable enemies of the United States. The new administration recognized that the Soviets and the Chinese separately sought some degree of American support in what had developed into a growing struggle against one other. This meant that international stability, including a degree of peace unknown since the beginning of the Cold War, might be achieved if the United States were willing to serve as a power broker in conflicts between the communist titans. From now on, Nixon and Kissinger concluded, American diplomacy had to be "triangular," a negotiation always involving all three nations—the United States, the U.S.S.R., and the People's Republic—conducted toward the goal of achieving a balance of power among them that would, in turn, secure world peace: a world peace in which the United States, economically more powerful and embodying the ideal of democracy, which Nixon and many others believed was an inherently more attractive political philosophy than communism, would nevertheless remain the dominant power.

To pave the way for the inauguration of an era of triangular diplomacy, military competition and political tension between the United States and the Soviet Union had to be reduced. To move toward this goal, Nixon and Kissinger espoused a policy of "détente" with the U.S.S.R., calling for an intensive round of diplomacy aimed at establishing an enduring accommodation. At the heart of the new policy were negotiations for a Strategic Arms Limitation Treaty (SALT), intended to slow down the nuclear-arms race and reduce the proliferation of weapons in the arsenals of both countries. The negotiations were opened in November 1969 and continued throughout most of the 1970s. Both sides readily agreed on a single underlying point: the United States and the Soviet Union each controlled enough nuclear weapons to annihilate each other many, many times. Somewhat less readily, both sides conceded that the cost of continuing to turn out, to man, and to maintain nuclear weapons systems was a serious drain on the national economies. There was, in short, sufficient common ground to push the talks forward through tortuous debates over precisely what weapons would be reduced and precisely how one side would verify the other side's reductions. In the end, SALT produced two milestone treaties, the Antiballistic Missile (ABM) Treaty (1972), which put a cap on defensive nuclear weapons, and the SALT I Treaty (the Interim Agreement on Certain Measures with Respect to the Limitation of Strategic Offensive Arms), which limited the offensive missiles. Additionally, the Nixon administration was able to negotiate breakthrough U.S.-Soviet companion trade and financial agreements. Nixon the cold warrior had become Nixon the peacemaker. It was, as old political hands breathlessly (and somewhat nervously) marveled, a miraculous transformation.

Détente with the Soviets was one leg of triangular diplomacy. The next was normalizing relations with the People's Republic of China. As early as 1963, John F. Kennedy had talked about doing just this, but he backed off from the attempt because of the certainty that political conservatives–foremost among them Richard M. Nixon himself–would assail his administration as being "soft on communism." Kennedy labored under the heavy burden bequeathed him by his father, Joseph P. Kennedy Sr., who, as U.S. ambassador to Great Britain on the eve of World War II, had championed Prime Minister Neville Chamberlain's ultimately infamous policy of "appeasement" toward Adolf Hitler. JFK was always fearful that critics and members of the press would pass a judgment of "like father, like son" and label him an appeaser. Now that he occupied the Oval Office, however, Richard Nixon recognized that his conservative cold warrior credentials would serve to enable him to make the very overtures to China that Kennedy could not have made without incurring severe censure and political damage.

Not that Nixon was entirely free from fear of rightwing criticism. He made his initial approach to the Communist Chinese in deep secrecy, but when Chinese officials responded favorably, Nixon quickly emerged with the stunning public announcement on June 10, 1971, that he was abrogating the twenty-one-year-old embargo on U.S. trade with China. In the meantime, Henry Kissinger was dispatched to conduct secret negotiations to prepare for a presidential state visit to Beijing. The subsequent public announcement of the visit was even more of a shock than the lifting of trade sanctions had been. As Nixon had feared, his traditional base, the right wing of the Republican Party, rose up in arms, condemning the Nixon administration for suddenly "surrendering to international communism." Nixon nevertheless toughed it out.

He saw that most Americans approved of his diplomatic overtures, and there was remarkably little outcry when, in October 1971, the United States reversed its twenty-year-old stance denying the People's Republic a United Nations seat. Not only was the People's Republic admitted to membership, the United States raised no objection to the expulsion of the non-Communist Nationalist Chinese (Taiwanese) delegation. When the president actually made his historic visit to China in February 1972, the American public by no means recoiled, but instead avidly watched television coverage of the occasion.

With the hindsight of history, it is tempting to ascribe even more to the 1972 state visit than what it actually meant at the time. Full formal U.S. recognition of the People's Republic of China would not come until 1979, during the administration of Jimmy Carter. Moreover, throughout the rest of the Nixon presidency, the United States blithely ignored communist objections and continued to furnish military assistance to Nationalist China. Nevertheless, the visit was a step toward normalization, and it made the worried Soviets that much more willing to negotiate with the Americans. Nixon and Kissinger refashioned the American political image so thoroughly that the United States no longer loomed as an adversary to the communist powers but as a diplomatic prize coveted by both Beijing and Moscow.

Finally, there was the issue of Vietnam. Whatever else triangular diplomacy might accomplish in promoting the long-term peace and prosperity of the world, Nixon and Kissinger most immediately wanted to cut North Vietnam loose from sponsorship by and military assistance from both the Soviet Union and China. The administration's goal was to be able to say to both the Soviets and the Chinese that favorable economic and diplomatic relations with the United States hinged on their renouncing support for North Vietnam. Furthermore, Nixon and Kissinger called on the Soviet leadership to broker positive peace negotiations between the United States and the North Vietnamese.

With this overview in mind, we can turn to how Kissinger and Nixon imaginatively ordered the world as they now saw it. Both men came to their jobs from a background of years of practical and theoretical experience in diplomacy. Both had come to adopt an assumption of realism rather than idealism. Kissinger understood that jockeying for tactical advantage was—realistically—the motive for many of the actions any nation took. The problem was that attaining a balance of power, in which adversary nations were essentially rendered equal in the military and other power they possessed, depended on their mutual ability to resist the temptation to jockey for tactical advantage. To discourage the continuation of a tactical arms race, Kissinger believed it necessary to recognize that national interests had to be measured in terms of mutual security as well as unilateral power. All sides had to reject parochial patriotic idealism in favor of a more pragmatic realism. That is, Americans, on their part, had to stop thinking in terms of such ideals as peace and freedom (a kind of *Pax* Americana) and begin thinking in terms of mutual security and the *balance* of power. On their side, the communists had to give up policies driven by ideology—the impulse to disseminate the Marxist revolution worldwide—and instead adopt the same realistic objectives of mutual security and the creation of a stabilizing balance of power.

It was not easy for Richard Nixon to abandon his own brand of idealism—his belief in the moral imperative that democratic capitalism must at all costs triumph over communism. As a veteran cold warrior, he had learned to see the world through a distinctly ideological lens. Nevertheless, he converted to realism by the time he was in the Oval Office. Claiming that it was above all else necessary "to put the survival of [the] nation first," Nixon expressed his conviction that the United States shared with China a "common interest" in survival and prosperity, which trumped any ideological differences between the nations. Turning his back on the policy that began with Woodrow Wilson—the mission to spread democracy abroad—Nixon instead advocated the creation of what he called a "structure of peace" with China and the Soviet Union, which would rest "on the hard concrete of common interests and mutual agreements, and not on the shifting sands of naïve sentimentality."

In order to define the specific goals and objectives of this new realism, Kissinger and Nixon devoted intensive study to planning and identifying the nation's capabilities and interests. They concluded that bureaucratization hindered thoughtful and agile statecraft, so they reformed the entire policymaking process by radically centralizing it in the White House—between themselves—and making it intensely secretive. Bureaucracy, Kissinger argued, paralyzed U.S. policymakers by widening the array of available policy options even as it limited the capacity of statesmen to choose among them. In effect, Nixon removed policymaking from the State Department and concentrated it in the National Security Council (NSC), his own inner circle of advisors. Smaller and more agile than the traditional cabinet department, the NSC could more readily plan, analyze, and review all foreign policy options, providing the data and the situational picture needed to make effective decisions. At least, that is what President Nixon always claimed. His many critics pointed out that bypassing the Department of State allowed the secretive president to keep foreign policy secret—and to do so mainly for the sake of secrecy itself. Be this as it may, small, specialized groups within the NSC, such as the Senior Review Group and the Defense Program Review Committee, reviewed policy, prepared contingency plans, and generally brought all foreign policy matters into the tight grasp of the White House. President Nixon boasted that he always knew where he was going in foreign affairs, unlike most past administrations, which had operated conventionally through the Department of State. The traditional system lacked a clear sense of purpose and State Department advisors could do little more than perpetually play catch-up, continually responding to crises rather than making the kind of policy that avoided crises in the first place.

Realism gave Nixon and Kissinger an acute understanding of global strategy, and it was on this that they built their negotiating approach in pursuit of détente with the Soviet Union. Their belief was that the Sino-Soviet split, Soviet-American strategic parity, and a growing power struggle within the Kremlin itself combined to create a situation in which reducing Cold War tensions was clearly in the interest of both nations. This impression is what prompted Nixon to signal to the Soviets his desire to negotiate. As he proclaimed in his first inaugural address, he saw the world emerging from a protracted "period of confrontation" and "entering an era of negotiation."

Realism made President Nixon eager to negotiate with the Soviets, but it also strictly circumscribed the areas that would be open to negotiation. Previous administrations had pursued agreements with the Soviets exclusively for the purpose of defusing crises or reducing dangerous tensions. In contrast, Nixon and Kissinger assumed that, in many areas, Soviet and American interests did indeed sharply differ and would always differ, so they decided to focus negotiation efforts only on areas where it was demonstrably in their mutual interest to do so. Whereas President Kennedy had sought personal contact with Soviet premier Nikita Khrushchev to reduce mistrust, Nixon once explained to French President Charles de Gaulle that he cared little about whether or not he and the Soviet leaders could "trust" one another. Instead of focusing on personalities, he wanted to concentrate exclusively on issues. "We should be hard and pragmatic in dealing with the Soviets," he recalled (in his 1978 memoir titled simply *RN*) telling President de Gaulle in 1969. "They knew what they wanted and we must know what we want." Indeed, détente depended entirely on a mutual recognition of power and national interest. "Power is a fact of international life," Nixon declared in a 1970 address to the United Nations. Negotiations had to take this fact into account. Personalities were strictly secondary to the fact of power. In December 1971, speaking to the Business Council, Henry

Kissinger elaborated on this. For years, he said, Americans had mistakenly assumed that tensions with the Soviets were the result of "mere personal misunderstandings and that the remedy for national differences is the development of interpersonal good will." Nonsense. Tensions were to be reduced through negotiation over power and other items of interest to both nations. It is a mistake, Kissinger told his Business Council audience, to confuse "foreign policy with psychotherapy. What we want to do is deal with concrete issues in our relationships."

In May 1972, the Moscow Summit convened, at which the United States and the Soviet Union signed the SALT I agreement and during which President Nixon reiterated how he intended to deal with the Soviets, telling his Russian hosts that an enduring agreement among superpowers could not be arrived at by "mushy sentimentality or by glossing over differences which exist. We can do it only by working out real problems in a concrete fashion, determined to place our common interests above our differences." It was "realism" applied to decision making. The president put the situation yet more bluntly to his cabinet and White House staff when he returned to Washington. SALT, he said, was the result of "fundamental shifts in the world balance of power," not "woolly headed idealism."

As the official Department of States history of foreign relations from 1969 to 1976 (http://www.state.gov/r/pa/ho/frus/nixon/i/21100.htm) explains, if realism was the basic assumption underlying the foreign policy strategy of the Nixon administration, triangular diplomacy and linkage were the bases for the tactics by which that strategy was to be carried out. Realist analysis of the world situation had revealed to Nixon and Kissinger that the new multipolar balance of power made triangular diplomacy the tactic of choice. Relations with one country would be employed as leverage to coax concessions from another. With Soviet-Chinese relations in crisis and the emergence of China as a formidable global power, President Nixon saw an opportunity to establish positive relations with the People's Republic that would serve as a diplomatic counterweight to the Soviet Union and, in turn, give the United States added leverage in obtaining concessions from Moscow. As Henry Kissinger had explained to Time magazine correspondents back in December 1970, the ongoing Sino-Soviet border conflict gave the Soviets an incentive to deal productively with the United States so as "to free their Western rear so that they can focus more on China." This being the case, the Nixon administration would "put out the word," Kissinger told the reporters, "that we are restudying the China question." The hint, Kissinger believed, would motivate the Soviets to move more positively than they had in the past. In addition, an ongoing dialogue with the Chinese would serve as what Kissinger called a "counterweight" against the Soviets. This was the essence of triangular diplomacy as it was soon employed in the overtures to China and the SALT negotiations with the Soviets.

Was the triangular approach effective?

Consider: No sooner did President Nixon announce in July 1971 that he would visit China, than Soviet Ambassador Anatoly Dobrynin informed Henry Kissinger that his nation was interested in moving ahead with a summit. The result was the historic Moscow Summit of 1972.

In conjunction with triangular diplomacy, the Nixon foreign policy employed what the president and his national security advisor called "linkage." This was a negotiation tactic that linked progress in one area of negotiation with the Soviet Union to progress in another, making the one dependent on the other, so that granting a given concession was linked to receiving certain concessions. As President Nixon explained in a letter to Secretary of Defense Melvin Laird, "the great issues are fundamentally interrelated," so that "we must seek to advance on a front at least broad enough to make clear that we see some relationship between political and military issues." Accordingly, the Soviets "should be brought to understand that they cannot expect to reap the benefits of cooperation in one area while seeking to take advantage of tension or confrontation elsewhere." The linkage concept would prove central to working toward several major international goals, including détente, strategic arms control, and normalizing relations in Berlin and in the Middle East, and, most compellingly, ending the Vietnam War. As Henry Kissinger put it in testimony before the Senate Foreign Relations Committee, the Nixon administration "sought to move forward across a broad range of issues so that progress in one area would add momentum to the progress of other areas."

Before and during the Moscow Summit, both Nixon and Kissinger remained adamant on linking progress on détente and SALT to a Soviet pledge to persuade North Vietnam to negotiate an end to the Vietnam War. As early as October 1969, in a meeting with Ambassador Dobrynin in the Oval Office, Nixon and Kissinger bluntly stated that, if the Soviet Union "did something in Vietnam" leading to a viable peace settlement, the United States "might do something dramatic to improve" U.S.-Soviet relations. Dobrynin accepted this positively, but just two months before the Moscow Summit, in March 1972, North Vietnam used Soviet-supplied weapons to launch a major offensive. President Nixon responded by threatening to abort the summit if Kremlin leaders did not apply the promised pressure to Hanoi to call off the offensive and commence genuine negotiations in good faith. The White House tapes record the president's phone conversation with Henry Kissinger on this occasion. Nixon told him that the best approach was to "keep kicking them [the Soviets] in the balls."

Yet Richard Nixon understood that diplomacy, whether triangular or conventional, could not rely solely on kicking one's adversary where it really hurt. Nor was it sufficient merely to promise to stop kicking if given what was asked for. The president realized that America's involvement in Vietnam was widely perceived in the world as an example of U.S. imperialism. Accordingly, he repeatedly hinted that Soviet cooperation in securing an acceptable peace would be linked to a redefinition of U.S. policy. What came to be called the Nixon Doctrine held that the United States would henceforward honor all treaty obligations, including the extension of its "nuclear shield" to all allies; however, the nation would act to defend its allies only in cases of invasion by a major power and, furthermore, only if extending such aid was perceived to be in the American interest. The defense of the Asian allies of the United States, the Nixon Doctrine held, had to become first and foremost the responsibility of the Asians. In the Vietnam War, this policy was carried out as "Vietnamization," the staged withdrawal of U.S. troops as U.S.-trained and U.S.-equipped South Vietnamese forces accepted more and more responsibility for their nation's defense. Even before he became president, in an October 1968 campaign speech, candidate Nixon had declared that the South Vietnamese should "fight their own battles."

Realist analysis of the world situation required a frank assessment of the meaning of the Vietnam quagmire. It was this: the capacity of the United States to control world affairs was declining in a situation created by a multipolar balance of power. This did not mean that America was edging toward impotence. Far from it. But the nation's declining capabilities had to be brought into line with its multitude of global commitments and interests. It was in this spirit that President Nixon spoke of America as moving from a position of "predominance to . . . partnership." Kissinger explained that the new direction of the many U.S. foreign aid and assistance programs would be to maintain American dominance in a "world order in which the United States does not have to carry the entire burden. This means relating individual countries to others in their regions and then relating them to the United States." Interdependence would be emphasized rather than a direct dependence on the United States; yet the United States would continue to occupy the central position in the new network of interdependence among nations receiving U.S. aid and assistance. In the former situation of a bipolar balance of power, it was essential for the United States to have, in effect, a multitude of client nations depending on it. In the new multipolar world, however, even a nation as powerful as the United States could not productively sustain all of these direct dependencies.

Developing interdependence among nations closely allied to the United States was a way of keeping America dominant without draining its resources. By no means did Nixon want the nation to withdraw into what he condemned as a "new isolationism." It was true that, even as the poles of international power proliferated, expanding from just two to many, the United States was, relative to the rise of a multitude of powers, in economic, military, and even psychological decline. A sluggish economy and the hopeless burden of Vietnam were taking their toll on the national psyche. Many political leaders advised a diplomatic retreat, as it were, a judicious narrowing of America's overseas obligations. Nixon and Kissinger were at pains to execute a foreign policy strategy that would keep the United States productively engaged in the world yet would ensure that the world did not unduly burden America's national resources. Maintaining a position of strength, especially military strength, was realistically essential to achieving détente, the SALT agreements, and normalized relations with China. This necessity also made Nixon wary about withdrawing from Vietnam in a manner that signaled surrender. "If they think we are weak," Nixon said of the North Vietnamese leaders, "they are going to pounce on us. If they think we are strong, they are going to deal with us."

It was a bind the Nixon administration never fully resolved, but did attempt to manage. While progress toward ending the Vietnam War was slow and fraught with the president's manipulations and even outright deceptions, the Vietnamization concept, driven by triangular diplomacy and the related principles of realism and linkage, did provide a way out, even if the ultimate inadequacy of South Vietnamese forces and government doomed Vietnam to unification under a communist regime. Détente, SALT, and the opening to China, all were real and important gains, which made a world long dominated by the doomsday specter of nuclear annihilation safer and more stable.

TAKEAWAY

BEFORE HE BECAME PRESIDENT, it was easy to brand Richard Nixon as an American chauvinist, willing to enforce decisions with the main strength of the mightiest nation on earth. Once he was in the White House, however, Nixon revealed a keen awareness of a global reality in which recourse to brute strength could no longer either drive decisions or enforce them. Unilateralism, as he saw it, had become a dead end. Instead, decision making had to acknowledge, address, and ultimately balance multiple poles of power, not in an effort to impose ideology on others but to reach mutually agreeable objectives, the most important of which was a global security that enabled all nations to achieve an acceptable level of influence and prosperity. Although Richard Milhous Nixon would betray a stunning ethical blindness, perhaps no American president has ever seen so clearly the global matrix across which national decisions are ramified, amplified, and compounded, sometimes sooner, sometimes later, and for better or for worse. He played global politics with all the gusto and aplomb of a RISK virtuoso.

CHAPTER 9

Reaganomics

Trickle-Down and Its Discontents

Previous page: *President Ronald Reagan speaking in Minneapolis, 1982.*

Reaganomics

IN THE DECEMBER 1849 ISSUE of *Fraser's Magazine*, the essayist, historian, and philosopher Thomas Carlyle branded the then-emerging study of economics the "dismal science." The label, though derogatory and unfair, is not without a certain instructive value: at the heart of economics is the principle of scarcity, the assumption that the world always and everywhere offers insufficient resources to satisfy everyone's wants. This being the case, civilization has developed an "economy," a matrix of exchange—to obtain something wanted, one must possess and relinquish something someone else wants.

In and of itself, the principle of universal scarcity is a pretty depressing proposition, yet economics is "dismal" in another way. The economy (that is, the matrix of exchange as well as the decisions involved in the exchanges) of any developed civilization is highly complex. When something goes wrong in the economy, when the principle of scarcity really hits home for a lot of people, when commerce stagnates, there is a mad scramble for a solution, a quick, simple, sovereign answer, some silver bullet that promises to return the economy to health. In reality, however, the economy presents such a complex matrix of decisions among so many individuals and involving so many variables—ranging from the weather (which affects food production, energy consumption, and transportation, among other things), to quirks of individual consumer behavior, to group dynamics on the very largest of scales—that no silver bullet can possibly hit its mark.

Still, we keep shooting.

In 1980, the Republican presidential candidate ran against the Democratic incumbent, Jimmy Carter, largely on an economic platform. The decade that had just ended saw a long decline in the American economy and in Americans' perception of that economy. Candidate Ronald Wilson Reagan repeatedly asked voters, "Are you better off than you were four years ago?" For the majority, the question was ludicrously rhetorical. Of *course* they weren't better off. Yet the economic decline in which the nation wallowed in that election year had begun not during the Carter presidency, but in the late 1960s under President Lyndon Johnson, as the Vietnam War drained the national coffers and the government began borrowing at an alarming rate. During the administration of Richard Nixon, which followed that of Johnson, a series of heavy-handed price controls and other government interventions seemed to deepen the economic malaise. The nation fell into the grip of something soon dubbed "stagflation," a vicious combination of recession—a stagnation of the economy—and inflation. It seemed like a very cruel joke, since by definition recession and inflation are opposites that should cancel one another out. As a song popular in the Great Depression ran, "Potatoes are cheaper / Tomatoes are cheaper / Now's the time to fall in love." The 1970s, however, demonstrated that this was not necessarily the case. When price inflation goes out of control, as it did in that decade, yet the growth of the economy slows or stops, unemployment rises, and recession sets in, even as prices remain high. Neither potatoes, tomatoes, nor anything else were cheaper, despite reduced demand.

The real hell of the economic dilemma is that, in a stagflated economy, government efforts to correct the stagnation usually worsen the inflation, while those aimed at coping with the inflation tend to deepen the stagnation. Stagflation characterized much of the period during which Nixon and his successor, Gerald Ford, occupied the White House. President Carter, in turn, inherited the Nixon-Ford economy, which was assaulted even further by a deepening energy crisis, as vital commodities—oil, especially—rose sharply in price. Under Ford, the Nixon price controls were abandoned. Under Carter, a system of federal oil reserves was created in an effort to blunt the effect of any future short-term oil shocks. The key word, however, was *future*. The oil reserve system did virtually nothing to relieve the short-term pressure. Carter also authorized a contraction of the money supply by the Federal Reserve (under its chairman, Paul Volcker), but this move came late in Carter's presidency—too late to produce a measurable impact by the time voters trudged off to the polls.

Arguably, by the time of the 1980 presidential campaign, stagflation was on the wane, but the electorate was still feeling the pain and President Carter was squarely in the voters' crosshairs.

Candidate Reagan worked hard to keep him there. He exuberantly criticized the Carter administration for promoting "big government," carelessly increasing government spending. Reagan promised that he would reduce the growth in government spending. He criticized Carter for high taxes and promised that he would reduce income tax rates. He criticized government over-regulation and promised that he would radically reduce it across the board. Finally, although the Federal Reserve under Carter had tightened the money supply, candidate Reagan called for further tightening. Thus Ronald Reagan proposed four major economic measures: decrease government spending, lower taxes, reduce government regulation, and control the money supply. These were the four major tenets of what was hailed as Reaganomics, the latest in an arsenal of silver bullets promoted as a sovereign remedy for America's economic woes.

Ronald Reagan's message found a responsive audience. In the 1930s, a nation fed up with Herbert Hoover's hands-off approach to the Great Depression–his refusal to use federal resources to provide direct relief to people and businesses–voted for the big government of FDR's New Deal. In 1980, feeling the effects of economic discontent linked to what the Republican candidate and others identified as "big government," many Americans were looking for someone to do what candidate Reagan promised he would do: get "big government off our backs." It was a philosophy Reagan had long held. While running for governor of California in 1966, he famously quipped that "Government is like a big baby–an alimentary canal with a big appetite at one end and no responsibility at the other." It was one of the great one-liners in the history of popular politics.

During the presidential campaign, Ronald Reagan was frequently criticized for making similar quips instead of presenting fully formed thoughts. While contesting for the Republican nomination, George H. W. Bush lambasted Reaganomics, especially the notion that the way to raise tax revenues was to cut taxes, as "voodoo economics." Yet the Reagan message on the national economy continued to get through, and it continued to strike a responsive chord.

In fact, not all of this was quip and voodoo. Reaganomics quickly acquired significant backing from some leading academic economists as a legitimate and overdue reaction against the entrenched assumptions that had long guided American fiscal policy. In the 1930s, the New Deal had ushered in an age of Keynesianism, an economic theory based on the work of the British economist John Maynard Keynes. Keynes promoted the creation of a "mixed economy," one in which the state as well as the private sector were seen as having legitimate and important roles, even in a capitalist economy. Analyzing the mechanism of the worldwide Great Depression, Keynes argued that government policies could be aggressively used to promote demand and therefore counter high unemployment and currency deflation. By reducing interest rates and using government money to invest heavily in infrastructure—that is, financing massive programs of public works—Keynes believed that the economy could be sufficiently stimulated to alleviate the depression. By definition, these measures called for the institution of "big government."

In its time, Keynesian economics was revolutionary because it broke with laissez-faire capitalism—the cherished doctrine that markets and the private sector in general operate most efficiently without any state intervention. Keynes believed that the great flaw in the laissez-faire approach, especially in regard to ending the Depression, was that it did not provide any automatic impetus toward full employment. The belief in this deficiency was central to his economic theory and the chief reason for government's interventionist role in the economy. In his 1980 campaign, Reagan did not attack Keynesian theory directly, but rather made an emotional appeal for a return to the free-enterprise principles that had dominated American economic policy before the Great Depression. Suffused with patriotism and old-fashioned values, this call rang true with many in the electorate. More remarkably, however, it also excited a coterie of academic economists and other professional advocates of the so-called supply-side economics movement, which had long been diametrically opposed to the basis of Keynesian economics and its notion of stimulating the economy by means of increasing demand. Supply-side advocates called instead for stimulating economic growth by creating incentives for people to produce goods and services on the "supply side" of the economic equation. Instead of creating demand through public works expenditure and the like, government's role in a supply-side approach is to reduce income tax and capital gains tax rates, to deregulate businesses (moving back toward laissez-faire), and generally to get out of the way of the private sector. The economic benefit of reducing taxes would be a stimulation of private investment, which in turn would yield higher productivity. This, proponents argued, would increase economic growth and ultimately lower costs for consumers. Indeed, enthusiasts argued that the size of economic growth created by supply-side policies would more than compensate for the short-term costs of cutting tax rates. Not only would cutting taxes pay for itself, it would ultimately increase government revenue, they argued.

While supply-side theory greatly bolstered Reagan's political campaign, critics of both supply-side economics and of Ronald Reagan argued that the proposed approach would deliver immediate benefits to wealthy investors and big business at the immediate and long-term expense of consumers, particularly lower-income individuals and families who relied heavily on government spending—especially those on welfare.

In the face of critics, some influential economic theorists were summoned to rally support for both Reagan and a supply-side policy. The economist Arthur Laffer had popularized a theory that government tax revenue could be maximized by setting tax rates at an optimum point on a curve, which is somewhere between a 0-percent tax rate and a 100-percent tax rate—neither of which extremes would generate government revenue. Supply-side advocates pointed out that Laffer's model showed how high tax rates lowered the incentive to produce, thereby reducing potential tax revenues despite the higher tax rates. This led to the stunningly counterintuitive assertion that lowering tax rates levied on the supply side—that is, lowering corporate taxes and capital gains taxes (taxes on investments)—would actually generate more tax revenue by shifting the optimum point to the lower end of the Laffer curve. It was this assumption that George H. W. Bush, during the 1980 Republican primaries, dismissed as voodoo economics.

But for voters about to make their decision, it had a magic to it—quite literally. Here was a way to lower taxes, thereby stimulating the economy, and yet do so at no public cost because the lower tax rates would actually increase tax revenue. As for the poor, whose welfare funding would suffer, it was claimed that they would actually benefit because the manna produced by a well-stimulated economy would "trickle down" to them. The phrase *trickle down* was almost certainly intended to mock supply-side economics and lampoon the assumptions that some of Reagan's supporters drew from the Laffer curve, yet many voters took it as a serious purported benefit of Reaganomics. By providing incentives to investors and big business, even at the expense of reducing government aid to the poor, the poor would nevertheless benefit along with the rich and everyone in between.

Looked at from our post-Reagan historical perspective, this seems either a gross over-simplification of what was necessarily a complex decision matrix, or simply evidence of that very complexity. The range and sheer quantity of decisions that contribute to an economy are so intricate as to defy common sense. Common sense says that increasing demand through government works will stimulate the economy by providing a market for what is produced. Common sense also says that reducing government's role in creating demand will reduce the market for what is produced. Moreover, common sense suggests that giving to the rich will not benefit the poor—or the middle class, for that matter. Yet—and this clearly seemed plausible to voters in 1980 as well as to the majority that reelected Ronald Reagan in 1984—perhaps the American economy is so complex that common sense is trumped by supply-side theory, the Laffer curve, and the concept of trickle-down.

Did Americans make the right decisions in 1980 and 1984? After deliberating for more than a quarter of a century, the jury has yet to come in, although many have reached their verdict. Some believe that Reaganomics—especially supply-side theory and the trickle-down promise—worked. Tax cuts during the 1980s, they say, did lead to an overall increase in government revenue because of the stronger economic growth the cuts enabled. Others, however, make reference to the parable of the emperor's new clothes, arguing that to believe cutting taxes would increase tax revenues was an egregious instance of magical thinking in 1980 and remains so today, whatever supporting theoretical pronouncements are marshaled and however warm and fuzzy the feeling the doctrine may have given some people.

If there is anything resembling a consensus among economists today, it goes something like this: it is naïve to believe that tax cuts will *completely* pay for themselves, but research suggests that tax cuts do contribute to an increase in government revenue and therefore do at least *help* to pay for themselves, because they demonstrably stimulate some increase in economic growth.

Consensus or not, some diehard proponents of Reaganomics continue to cling to the Laffer curve, arguing that, properly executed, tax rate cuts more than pay for themselves. They point to the fact that following the Reagan tax cuts, federal tax revenues did increase. Yet current supply-side critics (such as Jonathan Chait in *The New Republic*, September 10, 2007) counter that these increases were largely the result of Social Security Payroll Tax cuts, which had been scheduled before President Reagan took office. Moreover, a close examination reveals that while tax revenues as a whole increased during the Reagan administration, those revenues drawn directly from individual and corporate income tax actually fell as a percentage of the gross domestic product (GDP). Moreover, notwithstanding candidate Reagan's promises to reduce the size of government and government spending, the administration's monumental defense budgets resulted in unprecedented government expenditures, which, combined with the tax cuts, produced record-shattering budget deficits.

Still, it is possible to argue persuasively that the American economy performed better during the Reagan years than during the periods before and after the Reagan presidency. Real economic growth, which is economic growth adjusted for inflation, averaged 3.2 percent during the Reagan era. It was just 2.8 percent during the period of the Ford and Carter administrations and 2.1 percent during the period of the first President Bush administration and the two terms of Bill Clinton. Real median family income–income adjusted for inflation–increased by some $4,000 during the Reagan administration, having experienced zero growth during the Ford-Carter years and an actual loss of about $1,500 during the Bush-Clinton period. However, even proponents of Reaganomics admit that the savings rate fell precipitously during the Reagan era.

No one disputes the fact that funding for social programs was sharply cut during the Reagan administration, and it has proved almost impossible to demonstrate whether or not "trickle down" compensated for these cuts. While government spending actually increased under President Reagan, thanks to deficit spending on the military, it is undeniable that government regulation of the private sector was indeed reduced. However, some economists point out that a number of the most important deregulation initiatives were begun under Reagan's predecessor, Jimmy Carter, including the deregulation of airlines and truck transportation, as well as the beginning of deregulation of railroads and the communications, natural gas, and banking industries.

The record of Reaganomics is even more difficult to evaluate when one looks at the longer-term costs of trickle-down economics. Budget deficits—a product of the tax cuts and the massive increase in military budgets—as well as an increase in the U.S. trade deficit (as imports greatly exceeded exports) surely contributed to the Savings and Loan scandal and crisis that so blighted the economic outlook during the administration of George H. W. Bush, and also to the precipitous stock market crash of 1987. By the end of the 1980s, the United States was transformed from the world's greatest creditor to the world's largest debtor. Even Ronald Reagan spoke of this unnerving transformation as the "greatest disappointment" of his presidency.

The very real difficulty of evaluating the many and momentous economic decisions made during the era of Reaganomics is testament to the "dismal" complexity of economic systems. Add to this the suspicion many of President Reagan's critics hold—that Reaganomics was not actually a sincere effort to reform and revolutionize the American economy, producing benefit for all economic strata, but rather aimed at the traditional big-money supporters of the Republican Party (that is, investors and business)—and the picture becomes even darker, the decision matrix more obscure. Long after Ronald Reagan left office, his former Director of the Office of Management and Budget (OMB), David Stockman, confessed that the central piece of tax-cutting legislation during the Reagan years, the Kemp-Roth Tax Cut, was always what he characterized as "a Trojan horse," intended first and foremost to reduce top-bracket tax rates—that is, to immediately benefit the wealthiest of Americans.

Yet even arguing that the motivation behind Reaganomics was more political than economic does nothing to refute the theory behind supply-side economics and thus does not constitute a definitive judgment of failure of Reagan-era economic policies. A policy can have a cynical, anti-democratic (or at least anti-Populist) motive and still achieve laudable economic effects.

WILL WE EVER HAVE ANYTHING APPROACHING a definitive evaluation of Reaganomics? If the ongoing debate about both the short-term and long-term effects of FDR's New Deal is any indication, the answer is no. Proponents, critics, and those in between will continue to, in good faith, find sufficient evidence in the infinitely complex matrix of the American and world economies to support any number of views. In the end, or at least for the foreseeable future, the public response to the economic achievement of the presidency of Ronald Wilson Reagan has left the arena of theoretical prediction as well as empirical analysis, instead occupying the realm of mythology. While economists continue to argue about the effects of Reaganomics, many Americans cling to the magic of the man himself. Republicans, especially, invoke his name with a nostalgic reverence once reserved for the first standard bearer of the GOP, Abraham Lincoln. In the years since he left office—and even before he died in 2004—federal, state, and local governments have been prodigious in naming buildings, landmarks, portions of highways, streets, schools, sports facilities, parks, and a multitude of awards and honors after the fortieth president. In the year of his death, proposals were out forth to rename the Pentagon after him and to replace the portrait of Alexander Hamilton on the ten-dollar bill with that of Ronald Reagan, who might also be put in place of Franklin D. Roosevelt on the dime. As of mid-2008, they remain on the table. Unable to decide on how to evaluate the economic legacy of Ronald Reagan, the American people have nevertheless decided to canonize him.

TAKEAWAY

A SKILLFUL DECISION MAKER visualizes her moves, much like one who has mastered RISK visualizes the effect of invading an opponent's territory—or not. Yet some decisions are made in context so vast and complex—a matrix so dense—that it is not only virtually impossible to account for the entire matrix before making the move, it may also be almost equally difficult to determine the effect of the move even after it has been made. In these cases, the great temptation is to abandon rational evaluation of particular decisions and resort instead to elevating the decision maker to quasi-mythological status, creating about him or her a cult of personality that makes definitive evaluation on an objective basis highly unlikely. Admiration and gratitude are beautiful sentiments, but they do not excuse us from doing the hard thinking necessary to make big decisions in a complex environment.

CHAPTER 10

Blackout 2003

Lessons from the Power Grid

Previous page: *Electrical transmission tower.*

Blackout 2003

SATELLITES IN EARTH ORBIT "SAW" the whole thing, before and after. On the night of August 13, 2003, the great northeast swath of the North American continent was ablaze. And that is how it should have been. Since the late nineteenth century, with the introduction of electric lighting, civilization has been a 24-hour enterprise. From space, the continent normally appears phosphorescent, but by the evening of August 14, the blaze was all but completely extinguished. The Northeast was dark, which meant that 10 million people in the Canadian province of Ontario—about 30 percent of Canada's population—and 40 million people in the United States—some 15 to 20 percent of the nation's population—were without electric power.

By 4:13 that summer afternoon, 256 power plants had gone offline. In New York City, there were initial fears of a terrorist attack. The 9/11 attacks on the World Trade Center were, after all, scarcely two years old, and ascribing anything out of the ordinary to the work of terrorists had become something of a reflex. "People were freaking out," a young corporate intern interviewed by a *New York Times* reporter recalled. But word soon hit the streets that the power failure was technological rather than geopolitical in nature, and that revelation provided some relief, even as traffic snarled along streets whose signals had gone dark.

There were plenty of accidents and flaring tempers. Elevators stopped working, and on this seasonably hot summer afternoon, so did air conditioning. Those lucky enough not to have been caught in an elevator suddenly idled in one of Manhattan's high-rises began long walks down flight after flight of stairs in wells dimly illuminated—if at all—by battery-powered backup floodlights. For some, it was too much. The *Times* reported on one middle-aged woman who collapsed after descending many stairs in the darkened Met Life Building. Paramedics reached her and tried to resuscitate her when she stopped breathing. They called for an ambulance, but none were "to be found quickly on this afternoon of sudden chaos." In a corner of Café Centro, the woman lay for more than a half hour, growing cold as the paramedics continued to work on her. Someone was finally able to flag down an ambulance, but she was dead.

People tried to leave the island of Manhattan and get home to the outer boroughs, to Westchester, to suburban Long Island, to New Jersey, and to Connecticut. Nothing moved. Subways and trains ran on electricity, a commodity no longer available, and many commuters were trapped underground, stuck between stations. Commuter buses left from such places as midtown's cavernous Port Authority Bus Terminal, which was now dark. Police officers at the Port Authority blocked the entrances to the terminal, directing and sometimes shoving would-be commuters out.

"'Where are we supposed to go?" a *Times* reporter heard one man shout to an officer. "There are a hundred thousand people out there on the street."

The harried cop looked the man in the eye, pushing nose and lips uncomfortably close to the thwarted commuter's reddened face. "What do you want from me? I didn't cause this blackout. Now turn around and get out of the building."

On every street, small-time entrepreneurs moved into action. In some places, restaurants and shops emptied their dead freezers and sent employees onto sidewalks to hawk melting gourmet-chocolate ice cream bars at fire-sale prices. Street vendors suddenly pushed flashlights and batteries—where did they get them all?—the flashlights going for $25 a pop and batteries $10 each. In Manhattan's many hotels, guests could not reach their rooms. Even if they were permitted to walk up darkened stairwells, they would have found their electronic key cards useless.

The blackout caused at least nine deaths—six fatalities were attributed to the blackout in New York City, another three elsewhere. There was massive inconvenience, and there would be monetary cost—ultimately estimated at $6 billion. But most of all, there was fear—a fear that set in especially as the sun approached the horizon.

What would happen after dark?

Longtime New Yorkers recalled two previous major blackouts, one in 1965 and one in 1977. Many recollected the earlier of the two events with a certain nostalgia. People had pulled together back then, perfect strangers though they were, and the epidemic of looting and other crime many feared did not materialize. But the second blackout in 1977 was something else again. Thirty-one of the city's neighborhoods—especially the poorest neighborhoods—were swept by an epidemic of looting and vandalism. Along a five-block stretch of the principal commercial street of Crown Heights, Brooklyn, seventy-five stores were picked clean by looters. In the borough's tough Bushwick section, arsonists ran wild, ultimately destroying thirty-five blocks of Broadway, the commercial street dividing Bushwick from what was at the time the even more depressed Bedford-Stuyvesant neighborhood. In all, New York City firefighters responded to 1,037 fires, and 1,616 stores were looted, burned, or otherwise damaged in riots. Nearly 4,000 arrests were made, the prisoners deposited into hot, overcrowded cells or even consigned to the dark cellars of dark precinct buildings.

Which way would it go now: 1965 or 1977? Economically, the New York City of 2003 was a lot healthier than it had been in 1977, and Officer Marc Donella of the Manhattan North Task Force, a cop one *Times* reporter interviewed shortly before the sun set, was optimistic. "My feeling is there will be no disturbance tonight. This isn't '77. And listen, nobody's blowing the horn. They're all being very patient. Over on Lexington Avenue, there are civilians directing traffic." That phenomenon—ordinary citizens pitching in, helping out, nobody having asked them to do it—is what many veterans of the 1965 blackout remembered. And it was happening now, nearly forty years later, all over the city.

It proved a good omen. There was some looting and violence before power was generally restored throughout New York City in the early morning of August 15, but only a tiny fraction of what had happened in 1977. In other affected cities and towns throughout the Northeast, the impact was similar: not nearly as bad as it might have been.

Still, it was bad enough—not just on account of the $6 billion in losses, but because of the sense of civilization's vulnerability—the vision of it as little more than the most fragile of veneers.

If New Yorkers and other residents of U.S. and Canadian cities had for the most part acted with considerable maturity, generosity, and kindness during the crisis, government leaders were quick to point fingers in any direction, as long as it was away from themselves. Canada's prime minister Jean Chrétien blamed a combination of nature and utility company incompetence in New York State, claiming that the blackout had begun when lightning struck a power plant in northern New York, which triggered a "cascading" failure of the power grid emanating from the point of the strike. New York power officials were quick to retort that the problem had not originated in the United States at all, pointing out that there was no rain storm, let alone lightning activity, in the area where the lightning strike was supposed to have occurred, and that the local power plant in this area was one of the few that remained in operation throughout the entire blackout.

Still, the Canadians refused to surrender. The nation's defense minister, John McCallum, now pointed the finger at an outage in a nuclear generating plant in Pennsylvania. Pennsylvania authorities countered his claim, all the power plants in their state, nuclear and coal-fired, had been functioning normally. (Only some hours later did McCallum admit that his sources on this matter had been in error.)

On the U.S. side, New York governor George Pataki turned the tables, putting the blame on Canada. Authorities at the New York Independent System Operator—an association of independent power producers in the state—had told Pataki that it was "virtually certain" that the failure "had nothing to do with New York State." The New York ISO believed the initial failure had "occurred west of Ontario," from which it "cascaded . . . into Ontario, Canada, and through the northeast."

Sometime after Governor Pataki pointed north of the border, CNN reported that an unnamed official source had informed them that the Niagara-Mohawk power grid, serving much of New York State and parts of Canada, had been overloaded. This allegedly forced it to shut down, triggering a cascade of failures. The story was never confirmed.

Not that confirmation seemed to matter much, as people chattered in the dark. Everyone, it seems, felt entitled to an opinion. Even officials remote from the scene of the blackout were ready with definitive diagnoses. New Mexico governor Bill Richardson, who had been former president Bill Clinton's secretary of energy, pronounced the blackout the result of trying to power "a superpower with a third-world electricity grid."

The orgy of blame started within hours of the onset of the blackout, as technicians struggled to understand what had happened and how to fix it. It was only well into the night of August 14 that word began to emerge about the true location of the initial failure, which was somewhere in Ohio. At first, the analysis was indignantly denounced as rumor by Ohio's FirstEnergy Corporation, the regional supplier. Before the night was over, however, the president of NERC had an announcement to make.

The North American Electric Reliability Corporation (NERC) is an electric power-industry watchdog—a private-sector "self-regulatory" organization subject to audit by the U.S. Federal Energy Regulatory Commission and government authorities in Canada—whose mission is to improve the reliability and security of the bulk power system in North America. The NERC president confirmed that the problem had, in fact, originated in Ohio.

The designers of high-voltage electrical transmission lines take into account the fact that lines sag between the towers that carry them, and the hotter the lines become, the more they sag. Two factors raise line temperatures: ambient temperature, such as the heat of a summer afternoon, and increased electric load. The higher the current pumped through the wires, the hotter they become. To compensate for the sagging, transmission towers have to be tall enough and lines pulled tight enough to prevent the high-voltage conductors from coming into contact with anything, shorting out, and failing. The most likely object a sagging line might contact is an overgrown tree. That's why utility companies spend a lot of time and money inspecting lines and trimming trees.

Or, at least, they are supposed to.

The U.S.-Canada Power System Outage Task Force published its final report on the 2003 blackout in April 2004. The document chronicled a timeline of multiple contacts between transmission lines and overgrown trees starting at 2:02 PM EDT, when the line between the towns of Stuart and Atlanta, Ohio, touched a tree branch at Walton Hills, creating a short circuit and causing the line to fail. At 2:27 the Star-South Canton line also contacted a tree, tripping a circuit breaker, which reset after the line came free of the tree. Winds pushed the treetop into this line two more times by 3:41 and 33 seconds. Each momentary contact resulted in a short circuit, which tripped breakers that shut the line down, the breakers recycling on when the wind blew the swaying treetops free of the line. Two seconds after the third contact, however, the line contacted a treetop for a fourth time and was again shut down, but because the breakers had repeatedly cycled off and on in a brief period, they were automatically locked out to protect the lines, the breakers, and the generators from damage. The transmission load of the Star–South Canon line now dropped to zero.

The Outage Task Force report observed that the "line trips were not random. Rather, each was the result of a contact between a line and a tree that had grown so tall that, over a period of years, it encroached into the required clearance height for the line. As each line failed, its outage increased the loading on the remaining lines. . . . As each of the transmission lines failed, and power flows shifted to other transmission paths, voltages on the rest of FE's [FirstEnergy's] system degraded further . . ."

Thus the immediate cause of the biggest power outage in North American history was a few overgrown trees in Ohio, the tops of which made contact with power lines between 2:02 and 3:41 PM, shorting them out. Each short caused the affected line to either suffer a drop in its normal transmission voltage or to shut down altogether. In either case, the lost electrical load was automatically shifted to other lines. When this occurred, the temperature of the line to which the load was shifted rose, causing the line to sag. In the case of a line running between Hanna and Juniper, Ohio, the sag was sufficient to bring it into contact with an overgrown tree at 3:32, causing its failure.

Up to this point, the successive failures had been in the very high-voltage, long-distance transmission lines, each of which carries 345,000 volts (345 kV). Electric utility companies transmit power over long distances at high voltages because at these voltages the transmission of electrical current is most efficient, losing less energy to resistance. Of course, no desk lamp or electric toaster uses 345,000 volts. Before the current reaches consumers, it is stepped down by transformers at various points in transmission. The first step down is from 345,000 to 138,000 volts (138 kV) as lines approach distribution to urban areas. The Task Force report explained: "As each of FE's 345-kV lines in the Cleveland area tripped out, it increased loading [current] and decreased voltage on the underlying 138-kV system serving Cleveland and Akron, pushing those lines into overload." When voltage drops but current increases, the line becomes overloaded and hot, causing circuit breakers to trip in order to prevent damage to the line and the rest of the system. "Starting at 15:39 EDT," the Task Force report continued, "the first of an eventual sixteen 138-kV lines began to fail. . . . Relay [circuit breaker] data indicate that each of these lines eventually ground faulted, which indicates that it sagged low enough to contact something below the line. . . . As the 138-kV lines opened [that is, shut down], they blacked out customers in Akron and the areas west and south of the city, ultimately dropping about 600 MW [megawatts] of load."

The loss of high- and intermediate-voltage lines in Ohio triggered a succession of circuit breaker line trips. Many of the circuit breakers were of an older configuration, designed to respond to overloads—the condition created on one line by the failure of another—rather than to what the Task Force characterized as "true faults on the grid." In other words, the circuit breaker relays were of a design that did not "see" the true nature of the developing problem. For this reason, the tripping of these breakers exacerbated the crisis rather than alleviated it, creating a grid fault where there was none, spreading a local problem far and wide in a kind of "cascading" movement. The speed at which the circuit breakers tripped, according to the Task Force, "spread the reach and accelerated the spread of the cascade beyond the Cleveland-Akron area."

"A cascade is a dynamic phenomenon that cannot be stopped by human intervention once started. It occurs when there is a sequential tripping of numerous transmission lines and generators in a widening geographic area." The cascade can be triggered by "just a few initiating events," as in the case of the August 14, 2003, blackout. "Power swings and voltage fluctuations caused by these initial events can cause other lines to detect high currents and low voltages that appear to be faults, even if faults do not actually exist on those other lines." When this happens, generators are automatically shut down to protect them from severe power and voltage swings. At the same time, line relay systems work to isolate lines and generators from the rest of the grid in order to protect them. In the event of isolated problems, the system works well. "But when power system operating and design criteria are violated because several outages occur simultaneously, commonly used protective relays . . . cannot distinguish between the currents and voltages seen in a system cascade from those caused by a fault." The initial cause of the unanticipated multiple outages was distinctly low tech: a lot of untrimmed, overgrown trees. The shorts these caused overloaded lines, and that in turn led "to more and more lines and generators being tripped, widening the blackout area." Thus, a series of outages in northeast Ohio unleashed a cascade of interruptions that rapidly spread east and across much of the northeast United States and Canada.

Had the local utility pursued a rigorous program of tree trimming, there would have been no blackout because the immediate cause of the failure would have been eliminated. But there was more than this immediate cause.

At 12:15 on the afternoon of August 14, a little less than two hours before the first contact between a tree and a line, the software used to estimate and regulate voltages and loadings on the Ohio grid began producing what was clearly some erroneous data. This was the result of an outage in a line that had not been updated in the software. An analyst therefore manually updated the information and got the estimator software working properly again at 1:07. Having done this, he left for lunch. The problem was that, in order to troubleshoot the glitch, the analyst had had to turn off the trigger that automatically runs the estimator software every five minutes. Having fixed the problem by updating the data, he neglected to re-enable the automatic trigger before he left for lunch. At 1:31, while the automatic monitoring feature of the estimator software was still turned off and the analyst in charge was unavailable, a generating plant in the Cleveland suburb of Eastlake, Ohio, tripped offline during a period of high electrical demand. This placed a strain on high-voltage power lines, causing them to sag and bringing them dangerously close to several untrimmed and overgrown trees. By the time it was discovered that the estimator software's monitor was not operating automatically—the discovery came at about 2:40—the 345-kV line between Stuart and Atlanta, Ohio, had already failed after contacting a tree. The automatic software trigger was re-enabled too late to take into account this failure, and by the time the system could be again updated manually, other failures had begun. The software was not fully restored to proper and accurate automatic operation until 4:04, two minutes before the cascade—that phenomenon beyond human intervention—began. For this reason, the decision makers running Ohio's FirstEnergy Corporation plants were effectively rendered blind to the unfolding events.

Operating a power system control center requires nonstop decision making in order to produce a balance among the supply of power, the demand for power, and the capacity of available transmission lines. These decisions are made with the aid of computers, but they still require human intelligence, and both the computers and the human operators need a high degree of what engineers call "situational awareness." They need to know exactly what is happening in real time and at all times across the complex network of a modern power grid. This data collection is the function of advanced monitoring hardware and software. A deficiency in any of the hardware, software, or the human operators can easily create a situation in which even a few faults in the system—a succession of lines contacting trees, for example—can bring on a runaway cascade of failure.

One safeguard against the failure of a given power system's affecting the larger grid is a NERC regulation requiring operators to respond to local failures immediately. If they can find no solution within thirty minutes, the operators must either get more power locally or from other regions, or they must "shed load," cutting power to some of the areas they serve. Customers, naturally, are not pleased by being cut off, but this action is sometimes necessary to prevent a system-wide collapse. In a truly pressing emergency, operators do not even have the half-hour grace period, but are required simply to shed load immediately.

In the case of the Ohio utility company on August 14, monitoring software had been turned off and, along with it, the automatic alarms that should have warned operators of the multiple line failures. Yet the human operators are assumed to have the training and knowledge to function adequately even in the event of computer-based system failure. When they observed a sharp drop in voltage on the Ohio portion of the power grid at 3:17 PM, the FirstEnergy controllers should have swung into action by beginning to shed load. Instead, they did nothing, passively allowing another line to take the load that had been dropped by the first failed 345-kV line. As FirstEnergy controllers struggled to understand what now had become a rapid succession of line failures, they did not alert system controllers in neighboring states, nor did they at this point start shedding load until the problem could be sorted out. To be sure, this would have meant cutting off power to the entire city of Cleveland as the 138-kV lines began to fail, but doing so would probably have saved the Northeast power grid from collapse.

The U.S.-Canada Power System Outage Task Force was very hard on FirstEnergy, laying the blame for the blackout at the feet of controllers who did not follow standard operating procedure, and citing corporate management, which had failed to ensure that routine tree trimming line maintenance had been carried out. But the Task Force also called for more adequate regulation of the power industry generally, including the establishment and enforcement of more uniform quality and safety standards.

Other critics have gone beyond the scope of the Task Force report to conclude that the potential for the massive blackout was actually built into the design of the entire North American power grid. The power grid, these critics say, is over-centralized and therefore incapable of responding with sufficient speed and agility when something goes wrong. Writing on August 21, 2003, a few days after the blackout, Amory Lovins, a leading energy consultant, likened the modern North American power grid to a helicopter in contrast to an airplane: "A fixed-wing aircraft can glide to a safe landing without engines, but without instantaneous active control and a tail rotor, a helicopter drops like a stone."

Just as equipping a helicopter with a more powerful engine will not address its basic safety vulnerabilities, so merely adding generating capacity and transmission lines to the power grid does nothing to fix the system. Lovins pointed to the example of Ontario, hard hit by the blackout and unable to call on its new nuclear plants to provide the necessary power. The one thing nuclear plants cannot do is recover quickly from emergency shutdowns. They are difficult to restart. "They're the opposite of a plant providing power at peak times," Lovins observed, "guaranteed unavailable when most needed." As for adding lines, the paradoxical fact is that more wires tend to increase rather than reduce the chances of cascading failures. Instead of creating failsafe redundancy—a good line taking over from a failed one—more lines mean more opportunities for failure; moreover, additional lines also enable local energy suppliers to sell more of their power outside of their region when local demand is low. This adds to the already formidable complexity of the grid and introduces difficulties attendant on controlling yet more power.

Instead of adding capacity, it is far more efficient and safer to reduce consumption, not by cutting back on productivity or population, but by using energy more efficiently. This can be accomplished by consumers as well as by the producers of electrical appliances. Consumers can practice greater efficiency, and appliance makers can incorporate design features that use energy efficiently. Energy producers also need to develop and employ better monitoring systems as well as so-called "demand response" technology, by which certain appliances (such as power-hungry air conditioning systems) used by participating consumers automatically shut down for an interval of time during periods of unusually high demand.

Most innovative are proposals to decentralize power generation by allowing local energy producers to move on and off the grid as the situation requires. Even more radical is extreme decentralization, in which certain industrial customers operate microturbines to generate their own power, and homeowners rely increasingly on solar power. It is even possible to reverse the customary grid flow: when microturbine and solar users produce more energy than they need at a given time, they can sell it to the grid rather than draw from it.

AT THE HEART OF AVOIDING FUTURE BLACKOUTS—and averting a future in which catastrophic blackouts become frequent occurrences—is understanding the decision matrix of energy production, distribution, and use, whether that matrix is in the form of a vast power grid that must be managed with the most minute and supple attention, or whether it embraces approaches to energy that allow consumers to make more decisions about their own energy destiny on the grid, off the grid, and somewhere in between.

TAKEAWAY

THE NORTH AMERICAN POWER GRID is a decision matrix. Like the RISK game board, it is a system of interconnected nodes, in which a decision made at one node affects others. Many decisions are made at many nodes from moment to moment. A bad decision at even a single node—and any decision that fails to take into account the entire matrix is potentially a bad decision—can cascade throughout the entire matrix, bringing it down. The only way to defend against bad decisions as well as their potential for cascading catastrophe is to develop "situational awareness," an ability to accurately read matrix-wide effects from the perspective of your particular node. In effect, it is the ability to see from your own point of view beyond the restrictions imposed by your point of view. If you are running an electric power grid, that means having the best equipment available, up-to-date computer software and hardware, and the competence as well as the experience to know when, why, and how to override the automated systems. If you are responsible for making decisions in any other system—a government, a business, an organized endeavor of any kind—getting beyond your own point of view requires an open mind, a disciplined imagination, and a large capacity for empathy, the ability to see reality from the points of view of everyone else on the grid.

BIBLIOGRAPHY

Introduction

Axelrod, Alan. *Everything I Know about Business I Learned from Monopoly*. Philadelphia: Running Press, 2002.

Hopkirk, Peter. *The Great Game: The Struggle for Empire in Central Asia*. Tokyo: Kodansha International, 1992.

Neumann, John von, and Oskar Morgenstern. *Theory of Games and Economic Behavior*. 1944; reprint ed., Princeton, NJ: Princeton University Press, 1980.

Chapter 1: Visualizing the Decision

Honary, Ehsan. *Total Diplomacy: The Art of Winning RISK*. Charleston, SC: BookSurge, 2007.

March, James G. *Primer on Decision Making: How Decisions Happen*. New York: Free Press, 1994.

Plous, Scott. *The Psychology of Judgment and Decision Making*. New York: McGraw-Hill, 1993.

Chapter 2: Bismarck's Brain

Lafore, Laurence. *The Long Fuse: An Interpretation of the Origins of World War I*. Long Grove, IL: Waveland Press, 1997.

Taylor, A. J. P. *Bismarck: The Man and the Statesman*. New York: Vintage, 1967.

——.*The Struggle for Mastery in Europe: 1848–1918*. New York: Oxford University Press, 1980.

Williamson, D. G. *Bismarck and Germany 1862–1890*, 2d ed. London: Longman, 1998.

Chapter 3: The Story of Kudzu

Baldwin, Juanitta. *Kudzu in America.* Kodak, TN: Suntop, 2003.

Cope, Channing. *Front Porch Farmer.* Atlanta: Turner E. Smith, 1949.

Lembke, Janet. *Despicable Species: On Cowbirds, Kudzu, Hornworms, and Other Scourges.* Guilford, CT: The Lyons Press, 2001.

Chapter 4: Fireside Chat

Freidel, Frank. *Franklin D. Roosevelt: A Rendezvous with Destiny.* Boston: Little, Brown, 1990.

Leuchtenburg, William E. *Franklin D. Roosevelt and the New Deal: 1932–1940.* New York: Harper and Row, 1963.

Chapter 5: Deciding on the New Deal

Axelrod, Alan. *Nothing to Fear: Lessons in Leadership from FDR.* New York: Portfolio, 2003.

Grafton, John, ed. *Franklin Delano Roosevelt: Great Speeches.* New York: Dover, 1999.

Chapter 6: Island Hopping

Axelrod, Alan. *The Real History of World War II: A New Look at the Past.* New York: Sterling, 2008.

Costello, John. *The Pacific War.* New York: Harper Perennial, 1982.

Horner, David. *The Second World War, Vol. 1: The Pacific.* New York: Routledge, 2003.

Hoyt, Edwin P. *How They Won the War in the Pacific: Nimitz and His Admirals.* New York: The Lyons Press, 2000.

Van der Vat, Dan. *Pacific Campaign: The U.S.-Japanese Naval War, 1941–1945.* New York: Simon & Schuster, 1992.

Chapter 7: The FedEx Idea

Birla, Madan. *FedEx Delivers: How the World's Leading Shipping Company Keeps Innovating and Outperforming the Competition.* Hoboken, NJ: Wiley, 2005.

Frock, Roger. *Changing How the World Does Business: FedEx's Incredible Journey to Success—The Inside Story.* San Francisco: Berrett-Koehler, 2006.

Trimble, Vance. *Overnight Success: Federal Express and Frederick Smith, Its Renegade Creator.* New York: Crown, 1993.

Chapter 8: Triangular Diplomacy

Dallek, Robert. *Nixon and Kissinger: Partners in Power.* New York: HarperCollins, 2007.

Nixon, Richard M. *RN: The Memoirs of Richard Nixon.* New York: Simon & Schuster, 1990.

Suri, Jeremi. *Henry Kissinger and the American Century.* Cambridge, MA: Belknap Press, 2007.

——. *Power and Protest: Global Revolution and the Rise of Détente.* Cambridge, MA: Harvard University Press, 2005.

Chapter 9: Reaganomics

Lekachman, Robert. *Greed Is Not Enough: Reaganomics.* New York: Pantheon Books, 1982.

Niskanen, William A. *Reaganomics: An Insider's Account of the Policies and the People.* New York: Oxford University Press, 1988.

Sawhill, Charles R., and Isabel V. Hulten. *The Legacy of Reaganomics: Prospects for Long-Term Growth.* Lanham, MD: University Press of America, 1984.

Chapter 10: Blackout 2003

Collins, Glenn. "The Blackout of 2003; Flashlights Went Fast." *New York Times.* Aug. 15, 2003 http://query.nytimes.com/gst/fullpage.html?res =9D0DE0D91730F936A2575BC0A9659C8B63>.

Gottlieb, Martin, and James Glanz. "The Blackout of 2003: The Past; The Blackouts of '65 and '77 Became Defining Moments in the City's History." *New York Times.* Aug. 15, 2003 http://query.nytimes.com/gst/fullpage.html?res=9500E3DA1730F936A2575BC0A9659C8B63.

Rashbaum, William K. "The Blackout of 2003: New York Concerns; After Assuming the Worst, City Tries to Make the Best of It." *New York Times.* Aug. 15, 2003 http://query.nytimes.com/gst/fullpage.html?res=9506EEDA1730F936A2575BC0A9659C8B63&scp=20&sq=blackout+of+2003&st=nyt.

Tauzin, W. J., ed. *Blackout 2003: How Did It Happen and Why? Hearing Before the Committee on Energy and Commerce, U.S. House of Representatives.* Darby, PA: Diane Publishing Company, 2004.

U.S.-Canada Power System Outage Task Force. *Final Report on the August 14, 2003 Blackout in the United States and Canada: Causes and Recommendations.* Washington, D.C.: U.S. Department of Energy; Ottawa, ON: Natural Resources Canada, 2004.

Zernike, Kate, with Michelene Maynard. "The Blackout of 2003: Travel Woes; Rippling Across Continent, Chaos Strands Thousands." *New York Times.* Aug. 15, 2003 http://query.nytimes.com/gst/fullpage.html?res=9C0CE4D91730F936A2575BC0A9659C8B63.

INDEX

PICTURE CREDITS

Courtesy of Prints & Photographs Division, Library of Congress: Chapter 3: LC-DIG-ppmsca-009254; Chapter 5: LC-USW33-035391-C; Chapter 10: LC-DIG-fsac-1a35248

Courtesy of Wikimedia Commons: Chapter 2: 1890 Bismarcks Rueckritt; Chapter 7: Pigeon Messengers (Harper's Engraving); Chapter 9: President Reagan speaking in Minneapolis 1982

Courtesy The Granger Collection: Chapter 4: The Granger Collection

Courtesy the National Archives, Washington, D.C.: Chapter 6: 26-G-3183, Chapter 8: Nixon Library 2-25-72